The Networked Health-Relevant Factors for Office Buildings

Werner Seiferlein · Christine Kohlert
Editors

The Networked Health-Relevant Factors for Office Buildings

The Planned Health

Editors
Werner Seiferlein
TIM Technologie-Innovation-Management
Frankfurt/Main, Germany

Christine Kohlert
RBSGROUP part of Drees & Sommer
Munich, Germany

ISBN 978-3-030-22024-2   ISBN 978-3-030-22022-8   (eBook)
https://doi.org/10.1007/978-3-030-22022-8

Foreword

What message does the architecture of our new office building convey? Does it sufficiently represent the corporate philosophy? Does the refurbishment or renovation of an existing property improve its functionality? Decision-makers in companies are likely to be more driven by these and similar thoughts when planning their buildings than by the interaction of health-relevant factors and their influence on corporate success. Correct is: The closely networked health-related factors make a significant contribution to the success of a company. To a large extent, they determine whether employees feel comfortable at their workplace, are motivated and perform at their best.

The influences on health and well-being of employees are multifactorial: the design of the workspaces, the materials used, the indoor climate, the lighting, physical influences such as noise or psychological factors such as colours, contact possibilities with colleagues and, last but not least, the furnishing of the workplaces themselves all plays a role. And this means that by far not all factors influencing well-being and health in an office or administration building are covered, because internal processes as well as team dynamics and management influences must also be taken into account.

This makes it all the more important for various professionals to contribute their expertise during the planning, furnishing and commissioning of a building. It is only when these multifactorial influences are taken into account from a variety of perspectives—including the perspective of the architect, occupational safety expert, occupational physician, hygienist, occupational psychologist, industrial psychologist, organizational psychologist and occupational therapist. This is why TÜV Rheinland specialists from various specialist areas work together and provide interdisciplinary advice to clients. Only in this way is a holistic view of the building and its future users possible.

According to the statistics authority Eurostat, the average working life in Germany in 2015 was around 38 years—many years in which employees spend a large part of the day at work and are exposed to both positive and negative influences. If the factors that burden a building after it has been occupied, such as increased levels of pollutants due to the evaporation of building materials or inventory, predominate and are associated with unpleasant odours, the air-conditioning system brings more frustration than pleasure. And if the employees feel impaired by a lack of privacy, too few opportunities for rest and breaks, incorrect lighting, lack of ergonomics at the workplace or other environmental factors, the performance of the persons concerned drops. In the worst case, illnesses such as allergies, headaches, musculoskeletal disorders or psychological stress can lead to an increase in sick leave or employee turnover. This is associated with a well-known and widespread phenomenon in which a considerable number of employees in new or renovated buildings complain about health problems, the so-called sick building syndrome. The aforementioned factors not only have a negative impact on the health of employees and the economic success of the company, but also on the employer brand in general: the chances of winning highly sought-after high potentials and well-qualified managers are declining. In addition, there are costs for the improvements in and around the building in order to eliminate the negative effects on the health and well-being of the employees.

These are good reasons to already consider the networked health-relevant factors in office buildings in the planning phase. This book provides a comprehensive overview of the relevant influencing factors and bridges the gap to active implementation with the checklist.

Dr. Walter Dormagen
Business Unit Manager, Hazardous Substances,
Microbiology and Hygiene at TÜV Rhineland,
Cologne, Germany

Norbert Wieneke
Business Unit Manager, Health Management and
Occupational Safety at TÜV Rhineland,
Cologne, Germany

Preface

In the beginning, there was the thesis in the room "Health in an office building can be planned".

Just as a heating system, for example, can be designed on the basis of experience and rules, so health can also be planned in an office building.

This book is the realization of an idea, an idea that grew out of the task of constructing an office building for a manufacturer of pharmaceuticals. The question of the networked health-relevant factors for office buildings quickly arose.

Through numerous discussions with various specialists, with the designated authors and editors, a framework with the necessary factors for the topic collection gradually developed.

This framework was optimized and further refined in a joint workshop with the authors. In group work, the requirements were formed into a content structure depending on the chapters and then the texts were designed by the individual authors thanks to experience, practice and imagination. Particularly noteworthy here is the interdisciplinary cooperation, but also the joint, goal-oriented cooperation of "competitors" (it is probably one of the few books or text contributions in which, for example, several people from the furniture manufacturer business (designfunktion, Vitra and Kinnarps) work on a goal). The authors responsible met and/or consulted each other on topics where the coordination of the text content was helpful (e.g. the © Seiferlein 4 × Ls, the colours in theory and practice, the influence on furniture; see Chap. 9, Table 9.1).

As is often the case, this development pattern illustrates the necessary teamwork, working hand in hand for a common goal: "The planned health".

The target group of the book is a readership from different disciplines—building owners, architects, engineers, lawyers, psychologists, ergonomists, doctors, professors (e.g. economics, engineering technology), students, project managers, buyers, marketing staff, managing directors, project stakeholders, project developers, investors etc.—as well as different types of people (cf. Seiferlein, Woyczyk 2017, p. 101 ff.: Der Mensch) in different roles. In order to optimize communication and knowledge generation (ibid., p. 119 ff.: Knowledge Generation), a language understandable to all was used (ibid., p. 16, Fig. 1-1: "The Networked Triangle"; p. 29 ff. for creating user requirements' specific, functional, programming, etc.).

The book is a guide that can be applied chapter by chapter, depending on the interests of the reader. This guideline underlines the goal of providing application-oriented and practical support. Possible problems are dealt with and possible solutions are pointed out and illustrated with examples. Recommendations are given for companies, both for conversions and for new buildings.

It was important to us to bring the different disciplines together, and we enjoyed the exchange very much and it was extremely enriching. Health is an important issue. For companies, the health of their employees has the highest priority. With this book, we want to give an overview and a guideline to all interested in the topic.

Frankfurt/Main, Germany   Werner Seiferlein
Munich, Germany   Christine Kohlert

Introductory Remarks

What does health have to do with an administrative building?

This question, which is so important for us humans, is investigated in this book. The essential factors are determined and discussed, and the health-relevant aspects related to an administration building are discussed in detail. This approach generates the plannable and networked factors that form the basis for the design of an administration building.

Here the economic or commercial share should not be strongly included in the determination of results. However, in Germany there are more than 40 billion euros

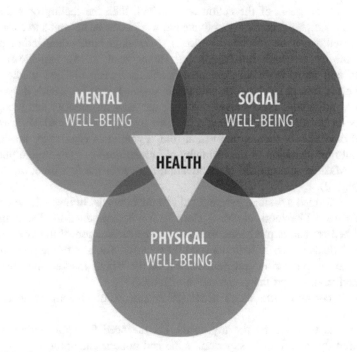

Fig. 1 Three levels of well-being

per year in lost production due to illness and inability to work caused by illness (cf. Spath et al. 2011, p. 40).

Health is a high good. It is therefore legitimate to work out the facets of health determinants and consider them in detail when creating a workplace for a certain number of people.

A book that meets this requirement creates a gap in the portfolio that is filled with it.

Particular emphasis is placed on the three areas defined by the World Health Organization as critical to health: mental–social–physical well-being in the workplace. All three levels play an essential role in human satisfaction with regard to work and the working environment (see Fig. 1).

Health Satisfaction

The statement: "Health instead of wealth" shows what is very important to Germans for their quality of life. Eighty per cent want health for family and partnership. Therefore, health care plays an important role as an interdisciplinary construct in which different experts, specialists, but also users should be involved.

Here, health is no longer measured in terms of illnesses, but in terms of satisfaction (cf. Martin 2006, Definition of job satisfaction If needs are satisfied in the work and/or the goals of the action are reached, then the feeling of satisfaction arises as a result. So someone is motivated to work because he wants to earn money and satisfied when he reaches his goal.). Nowadays, individualization plays a central role up to health satisfaction (cf. Huber et al. 2015). Holistic treatment concepts and individually implemented health care (personalized medicine) are goals for the future. Health orientation and responsibility for health lie primarily with each individual. Trends and drivers for the need for security and recognition and thus for health satisfaction were presented by Abraham Harold Maslow in his pyramid of needs for human motivation (the "Pyramid of Needs" by Maslow is a pyramidal representation of the hierarchy of human needs—in an article published in 1943, Maslow summarized his motivation theory for the first time; see Maslow 1943) (Fig. 2).

Maslow formed a 5-stage pyramid out of human needs. In Stage 1, he provided the basic need like food, clothing, sleep, warmth, reproduction. These must be fulfilled before man is motivated to strive for the satisfaction of the next stages.

Stage 2 deals primarily with the need for security. The aim of the programme is to promote the protection and demarcation of the labour market, the creation of stocks and savings and the preservation of jobs.

Stage 3 considers the social needs: love, affirmation, belonging to a group, human contacts.

Stage 4 is followed by the my needs, i.e. the need for appreciation, such as recognition by the group or superiors, fame and honour, attention.

The fifth and last stage finally consider the need for self-realization.

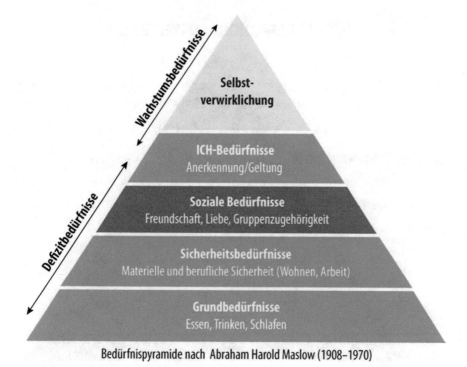

Fig. 2 Pyramid of needs, according to Maslow

Vischer (2008) developed this pyramid further to the personal satisfaction at work (Fig. 3). It looks at the psychological, functional and physical conditions under which people feel comfortable at work. In its pyramid, the lowest level of belonging and control is dedicated to physical well-being (correct chair, appropriate light, good acoustics). The next stage looks at functional well-being (air quality), and how the user is supported in his work. The highest level is concerned with psychological well-being, which is decisive for the quality of the workplace (feeling belonging and involved, having a home). This is also about safety, cleanliness and accessibility.

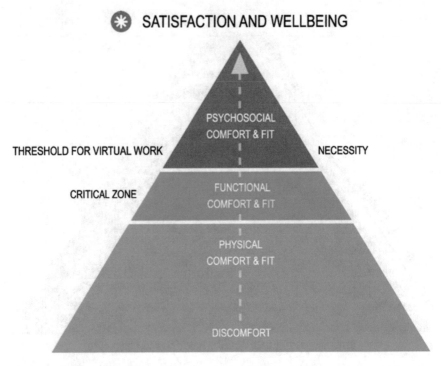

Fig. 3 Pyramid of well-being, according to Vischer (2008)

Healthy Despite Work

We spend at least a third of our lifetime as working days in the office. In principle, the design of office buildings and working environments is based on current standards and guidelines. The aim is to comply whenever possible with the European DIN standards and with the requirements of the BGI (Berufsgenossenschaftliche Informationen); see Sect. 10.2. However, guidelines can also be outdated and often lag behind the state of the art. Thus, for example, the workplace design can be have changed so that the directives and standards do exist and have done so for decades. The above-mentioned documents must be continuously revised by standards bodies, associations, companies or the legislator.

A higher standard is more and more often aimed at. These deviations are mostly based on company-specific documents with a separate standard and company-specific agreements (see Dietl 2015).

Designers always wonder what a design based on human dimensions and desires should look like. Various theories have been developed under the heading of "positive design". As an example, an example by Desmet and Pohlmeyer (2013) will be presented here, which is intended to promote human motivation by combining design for comfort, personal meaning and values (Fig. 4). In this context, design for comfort means reflecting on personal values, such as haptics, as well as on one's own significance and importance. Design for personal meaning means experiencing appreciation, being supported in achieving goals and enjoying friendships. Design for values refers to ecological design and sustainability. Apples, for example, stimulate people to pay attention to their own health. Rooms designed in this way help people to be motivated and have a lasting positive influence on their own well-being. The point is to be satisfied, to feel comfortable, to achieve something and to behave honourably in this environment.

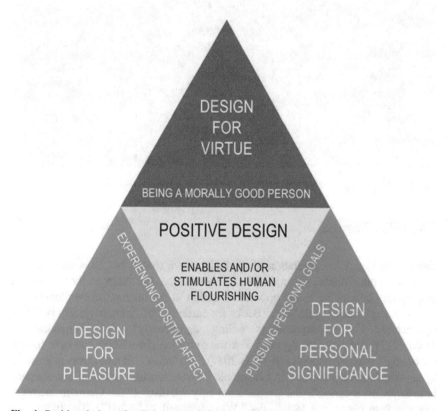

Fig. 4 Positive design, after Desmet and Pohlmeyer (2013)

A design that takes all three components into account supports human well-being. The happier we are, the more creative and successful we are (cf. Perlich et al. 2016: Gestaltungsdenken kann durch mobilen Raum fördern, Plätze, die sicher sind, zum Experimentieren; Mehta et al. 2012: People under moderate volume conditions create ideas that are more creative than those that were created either under low volume or high volume).

For the book, the health-relevant factors were investigated and arranged, which can affect humans in an administration building. These are far more factors than are addressed in legal standards and guidelines. There are no "guard rails" for these factors, and they can be planned and applied independently.

In the DNA structure of Fig. 5 are symbolically depicted the various factors for health. With this structure, the relation to knowledge and communication can be shown very well.

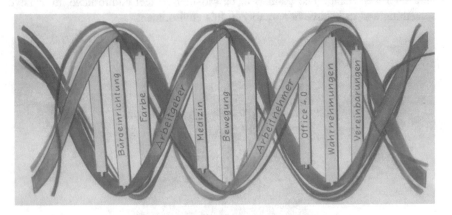

Fig. 5 DNA parable

The DNA Parable

The one in Fig. 5 the oscillation of the DNA curve shown can be compared with external social, political, cultural and other conditions. For example, the interests of employees often swing in the opposite direction to the interests of employers.

An example: Employers' order books are currently full, unemployment is falling significantly ("Unemployment is falling significantly", FAZ Wirtschaftsteil 31.03.2017), and sickness figures ("For the first time since 2006, sickness figures are falling", FAZ Wirtschaftsteil 31.03.2017). The "mood" of employers is growing again, which has a positive effect on employee-relevant measures and investments. In this "spendier mood", "the Germans are very satisfied" ("The Germans are more satisfied than ever since 1990", FAZ Wirtschaftsteil 17.03.2017). The trigger for this "once up and then down phase" depends on various reasons, such as reunification (1990), 11 September and Afghanistan war (2001), grand coalition (2004), beginning of the financial crisis (2008), Fukushima (2011), etc. (ibid.).

The "cross struts" does not of the DNA symbol "s" show the health-relevant parameters that have a significant influence on the employee's health and satisfaction.

The Generations

For whom are workspaces planned? For the old or the young employees? It is the top managers who answer this question with "for the boys". It is clear that the boys, statistically speaking, will use a new building for a longer period of time. But isn't it dangerous, in terms of appreciation and motivation, to concentrate on only one— the young—generation in the world of work?

How do the older generation feel about it? The company is also a place of business for those who, among other things, bring their experience and the knowledge they have acquired over the years as well as a certain composure to the company? Until when are you actually still young and from when are you already old?

One generation stands for about 15 years. The entry age into professional life is about 16 years and goes up to the age of 66, i.e. about 50 years. This means that at least four generations are working at the same time.

The industry's focus is on the very young. To the Generation Z belong 20-year-olds and younger. Generation Z is now entering the labour market (Needy 2016).

"Around three million people Z factors are already romping around on the labour market this year as a result of these considerations. For comparison: the predecessor generation Y there are about eight million members" (ibid.).

Today's young people work differently. "The mixture of work and private life is barely popular anymore. The young people of today have observed too closely how Generation Y often took their work home with them and could not get away from the laptop. The workers want regular working hours, unlimited contracts and clearly defined structures in their jobs," says Christian Scholz, an expert on the world of work: "When the working day is over, they don't read work e-mails". (ibid.)

Networking

Most industries have been able to improve the between administrative buildings, but also factory facilities and their internal and external links. The benefits of exchanging and using data are obvious. But how does this work, for example, in the pharmaceutical industry—in the regulated sector? Why does the healthcare industry find it so difficult to network data and information in the manufacture of pharmaceuticals? (Huber et al. 2015, p. 5)

The more intensively a company or plant is networked with its economic environment, the more favourable its innovative strength will be (cf. Spath et al. 2011, p. 14).

Essentially, it depends on the factors that can be networked.

Aim of the Book

The aim is to ensure that people working in productive and efficient work processes

- Find working conditions that are harmless, executable, bearable and free from any interference
- See standards of social adequacy met in terms of job content, job assignment, work environment, compensation and cooperation
- "Develop freedom to act, acquire abilities and be able to maintain and develop their personality in cooperation with others" (Schlick, Bruder, Luczak, quoted from Spath et al. 2014).

When planning an office building, the question arises: "What needs to be considered, what the employee values and demands"?

The focus of this book is on the factors well-being and health satisfaction (parameters such as innovation, future security were deliberately not addressed). The user requirements carried out in the early phase of numerous projects tend to articulate the demand for health in office buildings. One can therefore conclude from this that the employees latently demand this claim to health. This means that the well-being and health satisfaction are the focus of this paper—the focus is on plannable health. Other factors that can also be success factors—such as appealing architecture, transparency, clarity of objectives, strategic fit, risk management and value engineering, senior management support, planning responsibility and degree of individualization, project description and priorities, strategic orientation, alternatives, etc.—remain unresolved here (cf. Seiferlein 2005, p. 213 ff.).

Conclusion

With this book, the health-relevant factors are considered, which affect humans in relation to an office administration building and enjoy therefore in the planning special attention. Attached checklists simplify the upcoming planning work.

The book focuses on health and well-being. For this purpose, the factors that can be influenced by planning are identified so that the content of the book can be used on the basis of headings:

The Plannable Health

The following is a summary of the factors discussed in this book:

- Mutual agreement
- Individual perceptions (4 Fs)

- – Free air (and well-being)
- – Fuss flesh
- – Fair ("Because food and drink support body and Soul", Gutknecht 2008, p. 183)

- Colour in general and in particular
- Adequate office furnishings

  - – Office furniture for the workplace
  - – Furniture for the different areas (cooperation and exchange, concentration and recreation)
  - – Design of the offices ABW (office form)
  - – Ergonomics
  - – Guidelines, regulations (available)
  - – Laws, recommendations, warning signs

- Demand-oriented building services engineering
- Medical aspects

  - – Hygiene, allergies
  - – Worklife balance

- Movement at the workplace
- Outlook Office 4.0.

Literature

Dietl, A. (2017). Rules and regulations which the contractor does not know. In W. Seiferlein & R. Woyczyk (Ed.), *Projekterfolg - die vernetzten Faktoren von Investitionsprojekten*. Stuttgart: Fraunhofer, pp. 91–100.

Desmet, P. M. A., & Pohlmeyer, A. E. (2013). Positive design. An introduction to design for subjective well-being. *International Journal of Design*, 7/3, S. 5–19.

Gutknecht, C. (2008). *From Treppenwitz to Sauregurkenzeit. The craziest words in German.* Munich: C. H. Beck.

Huber, J., Kirig, A., Rauch, C., & Ehret, J. (2015). *The Philips Health Study. How trust becomes the driver of a new health culture.* Frankfurt/Main: Zukunftsinstitut GmbH.

Martin, P. (2006). *Mobile office work—Designing new forms of work humanely.* Düsseldorf: Hans Böckler Foundation.

Maslow, A. (1943). A Theory of Human Motivation. *Psychological Review, 50*, S. 370–396.

Mehta, R., Zhu, R., & Cheema, A. Is Noise Always Bad? Exploring the Effects of Ambient Noise on Creative Cognition. *Journal of Consumer Research*, 39/4, S. 784–799.

Needy, D. (2016). What generation Z expects from professional life. https://www.welt.de/wirtschaft/karriere/bildung/article152993066/Was-Generation-Z-vom-Berufsleben-erwartet.html. Accessed April 12, 2017.

o. V. (2017): Unemployment falls significantly. In FAZ business section, 31.3.2017. Accessed on ???.

o. V. (2017): For the first time since 2006, the sickness rate is falling. In FAZ business section 31.3.2017. Accessed on ???.

o. V. (2017): The Germans are more satisfied than ever since 1990. In FAZ Wirtschaftsteil 17.3.2017. Accessed on ???.

Perlich, A., Thienen, J. v., Wenzel, M., Meinel, C. (2015), Redesigning medical encounters with Tele-Board MED. In H. Plattner, C. Meinel, L. Leifer, (Ed.), *Design Thinking Research: Taking Breakthrough Innovation Home*. Springer Verlag, pp. 101–123.

Seiferlein, W. (2005). *Success factors in the early phases of investment projects*. Frankfurt/Main: Peter Lang.

Seiferlein, W., & Woyczyk, R. (Ed.) (2017): *Projekterfolg - die vernetzten Faktoren von Investitionsprojekten*. Stuttgart: Fraunhofer Verlag.

Spath, D., Bauer, W., & Braun, M. (2011): *Healthy and successful office work*. Berlin: Erich Schmidt.

Vischer, J. C. (2008). Towards an environmental psychology of workspace; how people are affected by environments for work. *Architectural Science Review*, 51/2, S. 97–108.

Werner Seiferlein
Christine Kohlert

Contents

Editors and Contributors

About the Editors

Prof. Dr.-Ing. Werner Seiferlein was born in Frankfurt/Main in 1960. He studied mechanical engineering at the Technical University of Darmstadt. He has been working in industry since 1985. During this time, he has worked as a production engineer and project manager for numerous investment projects at home and abroad with more than 90 branches worldwide (Russia, Poland, Ukraine, Egypt, India, USA, Japan, Brazil, Europe, etc.) and in the field of facility management. He gained his management experience as Head of Engineering Technology for Process Development, Industrial Engineering, Projects & Technology and Global Engineering Technology in pharmaceutical companies.

He was Senior Emergency Manager at Industriepark Höchst. In doing so, he assumed responsibility within the framework of the hazard prevention organization in accordance with the alarm and hazard prevention plan.

In June 2005, he was appointed to the TU Berlin, Faculty: Business and management, doctorate.

Since the winter semester 2006/2007, he has been Lecturer at the Frankfurt University for applied sciences in the field of economics for the compulsory subject "Introduction to Project Management". In 2013, he was appointed Honorary Professor.

He was Member of the Management Committee of ISPE Germany/Austria/Switzerland for more than 12 years. He is Member of various committees, such as the

Office 21 of the Fraunhofer Institute in Stuttgart and the DIN Committee in Berlin.

He was responsible for the master plan for rental and leasing of office space. He also coordinates FM activities at the Frankfurt Höchst site. Today, he is acting as Consultant (werner.seiferlein@TIMoffice.de).

Prof. Dr.-Ing. Christine Kohlert was born in Munich and studied at the Technical University of Munich. She is an Architect and Urban Planner as well worked as a Governmental Architect and has worked in various renowned offices. She is Managing Director of the RBSgroup in Munich and has been dealing with learning and working environments of the future for more than 30 years, in particular with the interaction of space and organization. Her focus is on the involvement of users in the development process, the visualization of the change process and spatial analysis. She attaches particular importance to the design of creative workshops and the active participation of future users through special tools to visualize the various processes.

She is also Professor at the Mediadesign University in Munich and the University of Applied Sciences in Augsburg. At MIT (Massachusetts Institute of Technology), she worked as Research Affiliate for 12 years on various research projects and conducted seminars on space and organization as well as innovation.

As Architect, she worked for renowned clients in the USA, Great Britain, China, Sweden and Eastern Europe. She lived for three years in Tanzania and one year in Kosovo and taught at the universities there. She worked for the Gesellschaft für technische Zusammenarbeit (GTZ), the Friedrich Ebert Foundation, Cultural Heritage without Borders (CHwB) and the German Embassy and was in charge of various urban development projects. She also supervised various development projects in Tanzania as part of the UNESCO curriculum and received her doctorate in European Urban Studies on the subject of port restructuring in Dar es Salaam at Bauhaus University in Weimar. She is active in various bodies and committees, research partner in Office 21 of the Fraunhofer Institute and the study "PräGeWelt" (PräGeWelt—Preventive Design of New Working Environments), part of the f.o.n. (flexible office network) and scientific advisory board of Euroforum.

Contributors

Peter Bachmann born in 1970, the environmental technician and marketing specialist was the initiator and project manager of the Sentinel Haus® research project from 2004 to 2006 (funded by the Federal Foundation for Environmental Affairs). Since 2007, he has been managing director and partner of Sentinel Haus Institut GmbH. The SHI currently has the most practical construction experience in the implementation of healthy buildings and is thus represented by Peter Bachmann, among others, as Expert in key political and institutional institutions and expert commissions. His focus is on practice and implementation, fundamentals and sales, law and marketing.

He is Founder and Managing Director of the Sentinel Haus Institute in Freiburg im Breisgau. In close cooperation with TÜV Rheinland, the company is successfully implementing its concept for healthier construction and refurbishment in practice throughout Germany. E-mail: bachmann@sentinel-haus.eu

Torsten Braun (Dipl.-Psychologe) was born in Bremen in 1958. After graduating from high school, he completed an apprenticeship as an industrial clerk.

From 1981, he studied psychology at the University of Trier with a focus on work, memory and perception psychology as well as experimental design. Already during his studies, he worked on research projects with the ergonomist Dr. Gerald Radl for IBM and also for Prof. Dr. Hartmut Wächter on the topic "Prospective work design in automation".

After 1988, he worked in lighting design. First, he worked as Project Manager at the lighting designer Christian Bartenbach in Innsbruck and then as Head of lighting applications at the luminaire manufacturer Zumtobel in Germany.

Since 1993, when the office "Die Lichtplaner" was founded, he has been employed worldwide as Lighting Designer. In addition to the administrative building, his work focuses on museum buildings and the implementation of light art works, including for James Turrell.

At present, the office has seven permanent employees. In addition to architectural lighting design, research projects on human-centric lighting in the office, perceptual behaviour in urban contexts or eye tracking movements and visual behaviour in tunnel structures are also supervised.

Since 2013, he is Lecturer at the TU Darmstadt in the field of design and building technology under the direction of Prof. Dipl.-Ing. Anett-Maud Joppien.

Karin Burghofer (Dipl.-Psych. Dr. phil.) was born in Siegenburg in 1970.

After graduating in psychology at the University of Regensburg (1989–1996) with a focus on communication in counselling, psychotherapy and management, she obtained her license to practise psychotherapy in 1999 and her doctorate in philosophy (Dr. phil.) in 2000.

From 1996 to the end of 2000, she was Research Assistant at the Clinic and Polyclinic for Surgery at the University of Regensburg. From 2001 to 2004, she worked as Research Assistant in the field of system analysis and process optimization at the Institute for Emergency Medicine and Medical Management of the Ludwig Maximilians University of Munich. In this institute, she was in charge of research and public relations from 2005 to 2013.

In 2009, she acquired a degree in behavioural therapy and has been Lecturer at the University of Health and Sport in Berlin since 2013. Since 2009, she has been working in her own practice in the Tegernseer Tal and in Bad Tölz as Psychological Psychotherapist and works in Munich as Consultant, Coach and Trainer.

She is Author of numerous books, chapters and scientific articles.

Dr. med. Michael Christmann was born in 1965, and he studied medicine in Mainz, Dublin and Lausanne.

During his many years of medical work with severely chronically ill patients, he decided to focus on preventive work in the future. Therefore, after qualifying as a specialist for internal medicine and the additional qualification in emergency medicine, he also acquired the additional qualifications in sports medicine, nutrition medicine DAEM/DGEM, preventive medicine DAPM and psychosomatic basic care.

In order to be able to reach people who are not yet very ill and therefore often do not go to the doctor at all, he acquired the qualification as Specialist for occupational medicine and worked from 2001 to 2008 in the company medical care of industrial park operators, chemical and automotive supplier companies.

From 2009, he worked in a preventive medical practice where, in addition to the early diagnosis of existing illnesses during check-ups, the focus is above all on promoting personal initiative by accompanying the individual on their way to finding a healthy lifestyle.

Since 2013, he has been Head of Health Management and Occupational Medicine Germany and Key Medical Doctor Central and Eastern Europe for a pharmaceutical company. It is committed to the further development of behavioural and relational prevention in line with key performance indicators in the awareness that some employees can be reached through individual behavioural optimization, but all through relational optimization.

Dipl.-Ing. Gerd Danner is Managing Partner of SoundComfort GmbH, which was founded in 2002 with acousticians from TU Berlin.

The planning office controls and coordinates projects for room acoustic planning and design of workplace situations. With the help of computer simulations, more than 150,000 workplaces have been planned acoustically in recent years and measures to improve office acoustics have been developed.

Birgit Fuchs Vitra GmbH, Key Account Management Seating/Ergonomics

- Studied humanities, M.A., and business administration
- Since 1996 in the office furniture industry,
- 2006 to 2015 consultant for ergonomics in the office

She was head of the ergonomics department at Vitra GmbH for ten years before she became Key Account Manager for large companies in 2015, advising them on change processes. They are always interested in the following questions: "What working conditions can I create to maintain the health of my employees?" and "What effect can the company achieve by changing the office space?

Dr. Rudolf Kötter was born in 1947, and he studied law (First State Examination in 1971) and economics (Dipl. Volkswirt 1975) in Erlangen in addition to philosophy and received his doctorate in philosophy in 1980. After many years at the Interdisciplinary Institute for Philosophy and History of Science of the Friedrich-Alexander-University Erlangen-Nuremberg (FAU), he was appointed Managing Director of the Central Institute for Applied Ethics and Science Communication of the FAU in 2005 and remained in this position until his retirement in 2015.

His working interests lie in the scientific theory of natural sciences and applied ethics, in particular economic and bioethics, and are documented in over 70 publications.

Thomas Kuk is True Sales Professional, with an early enthusiasm for design and furnishing. He was born in 1966, and he already knew during his apprenticeship with the leading Frankfurt office furnisher Emil Eckhardt Jr. that the furniture industry would be his passion. After a trainee programme, he becomes Junior Salesman there. In 1997, he moved to Objektform in Kronberg, where he successfully performed various tasks before becoming Managing Director Sales. Since 2003, he has been responsible for sales management at Spielmann Officehouse GmbH as authorized signatory.

Currently the company is part of the designfunktion group, Munich and Thomas Kuk is a member of the designfunktion management board as an authorized signatory.

Prof. Dr. med. Christian K. Lackner was born in Munich in 1961. After training in marketing, he studied human medicine in Budapest/Munich/L.A./San Francisco and Portland Oregon. This was followed by further training (surgery/accident surgery) and additional qualifications in emergency medicine and medical quality management. After his habilitation in acute medicine in 2003, he was appointed to the newly created professorship for emergency medicine and medical management at the LMU Munich and headed the associated institute at the University Hospital of Munich until October 2011.

Since April 2012, he has headed the Healthcare Division of the Drees & Sommer Group as Director together with a colleague. He has held a professorship in Berlin since October 2012.

His work and experience focuses on clinical and business management, risk management in clinical practice and the management of risk. (Acute) medicine and optimization of clinical and preclinical structural processes/patient treatment pathways as well as clinical business and organizational planning. In addition, clinical portfolio analyses (DRG), risk audits, needs analyses & projections, clinical change management/preclinical change management, quality management (CQI) of acute medicine/clinical practice. Medicine as well as human factor/risk management (acute) medicine, system implementations in acute medicine and the complex "space and health".

He is Author of numerous books, chapters and scientific articles.

Prof. Dr.-Ing. Ulrich Pfeiffenberger began his professional career in 1983 in an engineering office for technical building equipment, after studying mechanical engineering at the University of Stuttgart and obtaining his doctorate in engineering at the University of Essen. His main areas of work were air-conditioning and refrigeration technology. He was Project Manager for the entire technical equipment of medium to large projects.

In 1993, he founded the engineering company Pfeiffenberger with its headquarters in Neu-Isenburg. The office is primarily active in industrial construction with a focus on pharmaceutical production, refrigeration systems and comfort air-conditioning systems for special buildings.

In 1997, he was appointed as Professor at the Technical University Mittelhessen in Gießen for the field of Integrated Building Technology and Project Planning of Building Technology Systems.

He was Chairman of the board of the Fachverband Gebäudeklima for 15 years.

Dr. Katrin Trautwein who grew up in the USA was born in Stuttgart in 1962. After completing her studies in chemistry, she received her doctorate at the ETH in Zurich in 1991. Since 1996, she has been interested in colours and her research into Le Corbusier colours has made her famous. In 1998, she founded the colour manufactory kt.COLOR in Switzerland. On the heels of the Le Corbusier colours, other innovative colour ranges followed, including a white palette with 35 shades, a black rainbow and an ultramarine blue as deep as that of Yves Klein. She aroused the awareness for aesthetic and ecological coatings and aroused the fascination for the effect of natural pigments.

She is consulted as Lecturer and Consultant for historical colour palettes, colours and light contexts and colour concepts by universities, architects, preservationists and building owners. The German Werkbund honoured her work with the Werkbund label in 2014, she was invited by the Getty Centre in California as Scholar-in-Residence in April 2016, and her specialist seminars in Uster reach hundreds of architects and craftsmen every year. Her books *Black* (2014) and *225 colours* (2016) are available in stores.

Stephanie Wackernagel (Dipl.-Des. FH, M.Sc. Psych.) was recovered in Magdeburg in 1981. She studied industrial design at the University of Applied Sciences Magdeburg-Stendal and psychology at the University of Innsbruck (Austria). Since 2007, she has been advising numerous industrial customers of various sizes and from a variety of industries. She has worked for many years as a freelance product designer and for four years as Consultant and Senior Consultant for the RBSgroup Munich in the areas of change and communication management and analysis. There she led an internal research project funded by the Federal Ministry of Education and Research on the preventive design of new working environments. Today, she works as Research Assistant and Project Manager at the Fraunhofer IAO in the Competence Center "Information Work Innovation". She specializes in strategic consulting in the context of New Work Environments with a focus on change and communication management, scientific analysis and health promotion. She has been investigating the effect of the office environment on people's well-being since 2014.

David Wiechmann was born in 1971, and he holds a diploma in finance (FH) and studied journalism at the Institute for Journalism and Communication Research in Hanover from 1996 to 1999. After some editorial activities, he became Editor-in-Chief of the journal Mensch&Büro in the Konradin Media Group in 2006. In 2007, he also took over the management of the Mensch&Büro-Akademie, an academy that imparts specialist knowledge about office planning within the office furnishing industry. In 2012, he also became Editor-in-Chief of the international architecture and design magazine md Interior I Design I Architecture, which is also published by Konradin.

In 2009, he also developed a consulting concept for the implementation of corporate health management in companies for the Konradin Group. Parallel to his editorial tasks, he worked as Founder, Director and

Consultant for the Dr. Curt Haefner Institute in Heidelberg until he left the publishing industry.

Since 2015, he has been Head of Interior Design Team, Marketing Manager and Member of the Management Board in the German subsidiary of Kinnarps, currently the largest European manufacturer of office furniture solutions, based in Sweden.

He has been Honorary Board Member of the German Network Office since 2014. (DNB), which is committed to good and healthy office work as part of the New Quality of Work Initiative of the Federal Ministry of Labour and Social Affairs.

Abbreviations

ABW	Activity-based working
ADFC	General German Bicycle Club e.V.
BauPVO	European Construction Products Regulation
BGI	Berufsgenossenschaftliche Informationen (Information for the Employer's Liability Insurance Association)
CCL	Consulting Cum Laude
CRI	Colour rendering index
DGNB	German Sustainable Building Council
DIN	German Institute for Standardization e.V.
DNA	Deoxyribonucleic acid
FM	Facility Manager
HCL	Human-centric lighting
HVAC	Heating, ventilation, and air conditioning
I AM	International Conference on Harmonization
ILO	Institute for Industrial Engineering and Organization (Fraunhofer Institutes)
IT	Information technology
LED	Light-emitting diode
LEED	Leadership in Energy and Environmental Design
LOGI	Low glycemic and insulinemic
LRM	Clear room height
LUX	Unit of illuminance
MBO	Model building code
NWG	Not hazardous to water
PPD	Predicted percentage of dissatisfaction
R&D	Research and development
RCI	Repetitive strain injury
RTL	Ventilation technology
SVOC	Semi-volatile organic compound
TGA	Technical building equipment

TÜV	Technical Supervision Association
UBA	Federal Environment Agency
URS	User requirements' specification
VOC	Volatile organic compounds
WHO	World Health Organization

Chapter 1
Mutual Agreement

Stephanie Wackernagel and Christine Kohlert

1.1 People at the Centre of the Future World of Work

Powerful visions for a new world of work can herald sustainable and productivity-enhancing change for companies. After the introduction of a new world of work, however, many companies notice that they had too little courage in implementing their concepts. The expected opportunities have been fulfilled and many fears have not materialised. Often, dissatisfaction with a new office concept dissolves after a short time and opportunities for profound change are simply missed. In regular consulting projects at the Fraunhofer Institute for Industrial Engineering (IAO), it can be seen that project processes for the implementation of a new working environment are not determined by the vision of the company management, but by fears. The fears of decision-makers are fed by the fears of employees. These often have great concerns that their health may be in an open office structure. Overall, there are only a few scientific studies on the effect of modern office concepts on health. A current systematic literature search by the University of Freiburg (Lütke Lanfer and Pauls 2017) shows that the effects of open office structures on health are inconsistent and only of limited significance. Previous studies have not done justice to the complexity of an office environment. Nevertheless, one statement can be clearly stated: If open structures are designed, such as classical open-plan offices of the 70s to 90s of the last century, companies endanger the health of their employees (scientific studies on health in various forms of office): Bodin Danielsson and Bodin 2008; Bodin Danielsson 2013; De Croon et al. 2005; Oommen et al. 2008; Lucerne University of Applied Sciences and Arts, Seco 2010; Windlinger and Zäch 2007.

S. Wackernagel (✉)
Munich, Germany
e-mail: Stephanie.wackernagel@iao.fraunhofer.de

C. Kohlert
RBSGROUP Part of Drees & Sommer, Munich, Germany
e-mail: christine.kohlert@rbsgroup.eu

© Springer Nature Switzerland AG 2020
W. Seiferlein and C. Kohlert (eds.), *The Networked Health-Relevant Factors for Office Buildings*, https://doi.org/10.1007/978-3-030-22022-8_1

1

Of course, the working environment not only affects our health, but also our job satisfaction, our work performance and our work commitment (Windlinger 2017). Every company whose employees work in an office environment should therefore be interested in optimally shaping its future working environment. This chapter shows how the design of the office environment can promote people's health, how our satisfaction with office design is influenced and how companies can get their employees excited about the change to create a healthy working environment of tomorrow by creating a new working environment.

1.1.1 Health-Promoting Office Design

In an environmental-psychological study (Wackernagel 2017), three essential factors of office design were identified which have a positive influence on our health, our well-being and our satisfaction with office design: positive stimulation *by the room* (stimulation), coherence *of the* room (coherence) and *the room offers control* (control). The first factor stimulation is given when the office environment is generally attractive and varied and interesting additional areas complement the regular workstations. In particular, a positive effect comes from the office environment when interesting furniture is integrated into the working environment, coloured objects, interesting floor surfaces and unusual shapes of walls or wall elements. A multi-coloured design and the presence of plants also have a positive effect on the user. The study also shows that these design elements are still very poorly implemented in current office environments, but are more common in open office structures.

The second factor, coherence, describes the positive effect on well-being when the office environment is designed without contradictions. This is the case when it is clearly recognisable to the user how the spatial equipment (furnishing, furniture, etc.) is to be arranged[1], and the technical equipment (tools) and the interior equipment (doors, windows, etc.) are to be used. This also includes the fact that the user can clearly read the function of a piece of furniture and immediately recognise which function rooms or work areas have. This is possible if there is a clear difference between rooms or work areas (e.g. regular workstations, meeting areas, break areas, material storage, printer area, etc.). A harmonious office environment also exists if a structure is recognisable in the spatial design, there is a certain order and a "red thread" runs thematically through the design. This factor thus includes predictability of the experience we will make with the use of the space and its equipment. For this reason, it is also important that the user has sufficient space for spontaneous meetings and can orient himself well in his working environment. The design elements of the coherence factor are currently very well implemented in every form of office (individual offices, team offices or open structures).

---

[1] See Chap. 4 Adequate Office Equipment.

The third factor, control, describes the experience of security and being a gun in the office environment. This is what we feel when your workplace is protected from other people's eyes and others do not have the opportunity to look at your computer screen. Similarly, the control factor means that the majority of other workplaces cannot be seen at a glance from one's own desk. Furthermore, there is a positive effect on the user if workplaces are located at a pleasant distance from each other and the workplaces are visibly separated from each other (e.g. by fixed walls, partition walls or furniture). These characteristics meet our need for our own territory, which is strongly anchored in human evolutionary biology (cf. Altman 1970). In addition, people feel protected if not many people pass through the workplace, cannot approach themselves from behind and if spatial limitations of the workplace (such as fixed walls, partition walls, furniture, glass walls) provide sufficient protection against acoustic disturbances. But not only the direct protection at the own workplace plays a role here. In order to promote our well-being, it should be possible to withdraw for concentrated work or to switch off, as well as the possibility to have confidential conversations within the working environment. The investigation of various office types has shown that these protective elements are currently most strongly present in small-cell offices and only to a small extent in open office structures. A closer look at the individual elements reveals why small-cell structures are often preferred by the user and which concrete-adjusting screws can be used to meet people's need for protection with open office structures.

It is in the interest of organisations to offer employees' health-protecting and health-promoting working conditions and to create a prevention-oriented working environment. This strengthens the personal and social resources of the employees and thus maintains their health and performance. From a business perspective, this form of *ratio prevention* can reduce the costs of illnesses and absences (Ulich and Wülser 2015). A proven approach to the design of health-protecting working conditions is the involvement of employees (Slesina 2001). Through the collection and analysis of stress factors and requirements on the part of users, a preventive and prospective design of office environments can be achieved and, in addition, the willingness to change (the general willingness to change on the part of employees is related to procedural justice; McFarlin and Sweeney 1992) and acceptance on the part of users can be increased. In addition to research results from environmental psychology, such requirement analyses offer the planners of office buildings' essential adjustment screws in order to achieve a greater fit between the building and its users. It is not uncommon for users to explicitly dispense with design measures of the factor control in favour of quick coordination possibilities with colleagues and a more fluent workflow. Thus, the results of Wackernagel's study correspond to an ideal-typical design from the perspective of health promotion. Topics such as knowledge exchange, cooperation and work processes are not considered here.

1.1.2 The Perception and Evaluation of the Office Environment

The user is of course a component of the working environment . He interacts with the room and its interior design and thus not only perceives his office environment but also evaluates its possible uses (perceptions and evaluations can take place consciously or unconsciously in humans (Dijksterhuis et al. 2006)). Figure 1.1 illustrates that each user, each team or each workforce—shaped by the company's own work culture—assesses the possible uses of an office environment differently and that this office environment in turn is filled with life (behaviour) in a specific way by each user, each team and each workforce.

In practice, one repeatedly comes into contact with employees who nevertheless predominantly feel positively about an office environment that is obviously designed to be as stressful as it is lived to be stressful. In a case study, a company has an office environment that promotes communication and a team of eighteen employees worked in an open office structure. This included ergonomic and modern furniture and an attractive interior. Apart from two large meeting rooms, there were no additional work options, so that the employees carried out various tasks in the working room. During the inspection of this workroom, some employees worked on components, others on work contents that required a high level of concentration, other employees there conducted longer informal coordination sessions directly at the workplace and others made telephone calls in loudspeaker mode. In addition, a coffee machine was located in the room, which was in regular (loud) use. These mutual disturbances are unacceptable from a labour science perspective. In the dialogue with the employees, the office was nevertheless described as very positive. Only when asked whether they could perform each of these sub-activities well was it stated by the employees that retreat possibilities would be desirable. They quickly added, however, that they otherwise enjoy working in this office environment. How is that possible? In conversation with the team, it became clear that the colleagues enjoy working together very much. They appreciated the ability to quickly coordinate rather than edit certain work content without interruption. The evaluation of the office environment plays an important role in determining whether users are satisfied with it. Satisfaction is also the degree to which the user has the impression that the office environment serves him in achieving his goals (Canter and Rees 1982).

However, the first evaluation of a new office environment occurs long before a user comes into contact with it and can evaluate the concrete use via interaction. contentment or dissatisfaction with a new working environment, employees already

Fig. 1.1 Scheme of action between objective environmental characteristics and behaviour

feel the first information about a planned redesign. It is perfectly normal for changes to be assessed before a tangible picture emerges. The employees have explicit or implicit knowledge about comparable working environments or good or bad experiences with (supposedly) similar office structures. Each user is therefore subject to individual standards for the evaluation of a new office environment. We compare and evaluate the new with people in relation to what we know or trust. The familiar is perceived as positive and confirming. In the event of a slight deviation from the familiar, we are prepared to review our standards (cf. Nasar 1994).

A practical example from the determination of requirements for a new office environment for users of team offices (rooms with two to seven workstations) is described as follows: When employees "smell" the planning of an open office structure, the following situation could be experienced across many different companies. Each of this room situation is inadequate without alternatives or retreat areas. The users of team offices indicate that they can imagine working with max. one other person in the office. Users working in a 2-person office indicate that they can work in a 3-person office. Users from a 3-person office say that a maximum of a 4-person office is possible. In future, users from a 4-person office could have a maximum of five people in one room work, etc. Each of these room situations The surface of the vehicle is inadequate without alternative or retreat surfaces. Over the years, many companies have originally created small offices due to lack of space with additional staff. This led to the experience on the part of the employees that the work situation deteriorated with each additional person. Furthermore, an increasing effect is assumed: With each additional person, the potential for interference multiplies. If two other people in the room are already disturbing, how will it be with ten or twenty people? Users can seldom imagine a completely different way of use, which requires working in an open office environment, since neither experience nor knowledge is available.

1.1.3 The Willingness of People to Change

We trust in what has proven itself over years, possibly even over decades (Landes and Steiner 2014). If one's own work tasks have been successfully completed over a long period of time, e.g. in a cell structure, it may initially be completely incomprehensible to the employee why his company plans to change this ideal work situation. In rooms or room structures that we are used to, we humans feel safe and strong. An employee knows how to use the familiar office, which behaviours cause which reactions and has developed strategies that enable him to deal successfully with stress factors. A frequently cited example in practice is the closing of the office door as a signal for "Please do not disturb" and the creation of peace. This proven strategy is not applicable in an open office structure. Since losses weigh more heavily than profits bring joy to people (Samuelson and Zeckhauser 1988), it is not easy for many users to abandon proven strategies, to develop a willingness to learn new strategies and to discover the benefits of change.

These remarks could suggest that changes are particularly difficult for older employees. In fact, the willingness of older employees to change is estimated to be significantly lower than that of younger colleagues (Klinger et al. 2014). Resistance to change is not related to age. An extensive study by the University of Münster with more than 40,000 data records found that resistance to change is actually related to the time an employee spent in the same workplace (Hertel 2013)—an indication that the number of changes experienced influences our willingness to change. This is also the experience in practice. Resistance to planned changes comes increasingly from employees who have been with the company for a very long time and have experienced only a few changes during this period. But is the key to the successful realisation of a new working environment really the employee who is generally willing to change?

The assumption suggests itself that employees who are generally willing to change are open to a new office structure. A scientific study (Szebel 2015) examined the willingness of employees to change when introducing a new office environment in an energy company. However, employees who are generally willing to change are not necessarily open to the introduction of a new world of work. The project-related willingness to change is influenced by the perception of leadership, project communication and benefits. So how can organisations successfully accompany their employees into a new working environment (the following elements of a successful change management strategy are based on Kotter 2012)?

1.1.4 Successful Change Management for the Realisation of a New Working Environment

The Question of Urgency

There are many triggers for a necessary corporate change (Lauer 2014, p. 13 ff.). External factors and environmental dynamics such as demographic change, financial market, digitisation lack of skilled workers or climate change can lead to changes in the organisation of the internal factors such as strong growth, innovation requirements, strategic reorientation, building refurbishment, restructuring or processes alignment. Employees do not resist modernising the office environment, but why does the company also need a new office structure? Companies have to answer the question of urgency to their employees: If everything is going well, why should employees support change? If the answer to this question is not comprehensible and authentic, resistance is on a broad basis.

Showing a Powerful Vision

A change imposed by the leadership can have negative effects on the willingness to change and lead to a drop in performance in the sense of an internal dismissal or to an actual dismissal of employees (ibid., p. 58). A powerful vision, on the other hand, can inspire employees to actively support change. To allow employees for the planned change decision-makers must demonstrate a clear and easily understandable vision. Attractive goals motivate employees to get involved in the change process. A sound and credible strategy to achieve the goals set reaffirms the determination of management and motivates employees to support the change process. In practice, planning and consulting companies are often called in for a "relocation project" or "renovation project". There is seldom any sign of a concrete and attractive vision of the future. It is important that decision-makers take the time to develop a common vision in the management committee. This must be fully supported at community level so that company-specific objectives can serve as guidelines for all subsequent measures.

A Strong Leadership Coalition

The management must be an active supporter of change and visibly stand behind the change for the employee. The realisation of a new world of work requires leadership, not just good management. A strong leadership coalition must embody the vision vividly. If the leadership shows no interest in the change, why should the employees? In a case study, the modern workplace concept of a parent company was to be implemented at a subsidiary company. The subsidiary had no influence on the concept and had to implement it almost one-to-one. The management of the subsidiary was not very enthusiastic about this and let its employees know. This attitude was received as a signal of solidarity and certainly served the harmony between company management and employees. This attitude did not contribute to employee support. The resistance attitude was transmitted on a broad basis and the statement "those above do not like it either" was a regular reason for the lack of willingness to deal with the opportunities. Consistent supporters of the vision of the future even exemplify the propagated goals. Company managers who themselves work in an open area during the introduction of an open office structure actively motivate their employees to support change and make the vision a living reality together.

The Emotional Roller Coaster Ride in the Change Process

The emotional management of a change is reflected in the frequently cited change curve (Landes and Steiner 2014, p. 14). This process is also reflected in projects for the planning of a new office environment. In the status quo, the majority of the workforce is satisfied with the familiar office situation and is working at the usual level of productivity. When the announcement of a change in the existing office structures is made, employees can be worried about and fear react. It is the indeterminate feeling of fear of something new that could overtax the individual. With the announcement of something new, the tried and tested is inevitably attacked and can indirectly convey to the employee that something is "wrong" so far. Personal insecurity, fears of difficulties and worries about increased workload cause stress among those affected (Lazarus and Folkmann 1984). The early provision of information on the project and its progress counteracts the employees' feelings of insecurity. The question of urgency should therefore be answered at the start of the project and the past achievements explicitly acknowledged. In practice, the first information event is held at this time in the best case, but often only an info mail is sent by the company management with the first rough facts. In this phase, many project managers leave out critical topics for the time being. Information is deliberately communicated that is very vague, so that employees cannot yet derive anything concrete from it. Company decision-makers do not want to bring unrest into the company and only communicate concrete steps when decisions have been made. Very few change consultants believe at this point that it is precisely this behaviour that fuels the rumour mill.

Vague formulation of objectives The employee can be perceived as an indirect attack on the person if he or she cannot grasp what is exactly wrong with the current work situation and in what way he or she and his or her performance relate to it. Once again an example from practice is described as follows: Many company managers hope that a redesign of the office environment will lead to a reduction of knowledge islands in the company and a higher level of cooperative exchange between employees. Behind this are often the goals of streamlining the work process and the qualitative optimisation of work results. In fact, promoting face-to-face communication within the company can lead to employees receiving more important information and developing significantly more ideas and new solutions (Rief 2017). However, employees are often only passed on the vague goal of a working environment that promotes communication. In a case study, one employee rightly argued that the formulation of the goal of "promoting communication" was not the right way to go about it. In the "management" could assume that the employees have not yet communicated and thus do not perform their work well. Goals must become as tangible as possible for employees in order to counteract resistance.

Resistance on the part of the employees can express itself in very different forms such as reproaches, restlessness, evasion or listlessness (more on forms of resistance: Doppler and Lauterburg 2014, p. 339). In this phase, the productivity of employees can even increase slightly to provide counter-evidence that everything is working perfectly at the moment. The quicker the employees become aware that change will come irreversibly, the sooner a rational acceptance of the impending change can take place. This is often associated with the emotional experience of frustration, depression and a drop in productivity. Reality is accepted and the process of letting go can take place. Here, too, the appreciation of jointly experienced successes plays a strong role. The old building and the old structures are said goodbye in gratitude. Joint activities can support this process. In a case study, the old building of a company should be completely renovated. Many walls that had to be demolished for the renovation offered the opportunity to do justice to the dubbed "demolition party". Employees could symbolically support the demolition. Spray cans and sledgehammers were ready to tear down the old (thought) building and have fun together saying goodbye to the old one.

Dealing with the new working environment is a longer process that can be supported by various measures over the course of the project. There is no prototype change management programme. There are proven elements, some of which are listed below, but each measure should be specifically adapted to the corporate culture. In one case study, an exceptionally large number of measures were implemented, but change issues were dealt with exclusively on a rational level. In case studies in which an emotional confrontation with the future working environment is allowed, e.g. by explicitly allowing feelings of anger in workshops, the change process can also be successfully shaped with fewer but targeted measures.

The change management roadmap should also be adapted to the different target groups in the project. "Everything for everyone" is not helping. Different groups of employees can deal with the future in different ways and at different times. A regularly used change barometer enables the change management team to respond to the current information needs and moods of the workforce. This is again best achieved through direct dialogue. *Coffee talks* or *regulars' tables* can be offered regularly for voluntary participation. In practice, these take place separately according to target group. An informal exchange opportunity for managers supports open dialogue about the same challenges and how to deal with them. Managers should offer their employees security in the change process. Of course, they can openly admit in front of employees that the changeover can be difficult for them too, but they should always give the employees confidence. Leaders increasingly develop confidence themselves by addressing their own uncertainties and concerns. A regular opportunity for exchange between user representatives (change agents) also makes it possible to discuss similar challenges. Accompanied by a change manager, this form of exchange can regularly be used to keep an ear on the staff and the change management timetable can be adapted with targeted measures in line with requirements.

Successful Project Communication
The introduction of a new working environment must be based on information and communication measures for the entire workforce (Kavanagh and Ashkanasy 2006). The more comprehensive the information offered on the project and the more extensively employees feel involved in the project, the more willing they are to change (Schweiger and DeNisi 1991; Larsson and Lubatkin 2001). Knowledge of the impending change is needed so that employees can change at all (cf. Szebel 2015, p. 109) information . The employees should address both the rational and the emotional levels.

Rational information provides high-quality information on the new work environment, the project and the concrete change process. The more positively the project and the future working environment are perceived by the user, the greater his willingness to change (ibid., p. 107). This does not happen, however, when the employee is exclusively presented with the advantages of the project. The information must address the personal benefit and be experienced as personally relevant. In addition to the opportunities offered by change, it is therefore also advisable to address negative aspects—the price of change. If companies represent only the positive aspects, employees can either not feel taken seriously or become suspicious, these feelings favour resistance. Employees must also understand the urgency of the change mentioned above. To this end, the individual user must become aware of the need for change on his side. Another element of communication is giving security. Not everything will change for the employees in future world of work. Many things remain the same and as stated above gives us familiar security. In order to find out what opportunities your own employees see in the new office concept, what fears they have, what gives employees the most security despite change and what urgency employees see for change, direct dialogue with employees is indispensable.

The courage to face even uncomfortable topics gives the company the opportunity to meet employees via a tailor-made communication system. In practice, change managers reflect these four fields of change—success factors , price of change, security and urgency—in workshops, e.g. with employee representatives and managers. By dealing with the various aspects, the participants can penetrate the project more deeply and work on their own position on it. It is not only the project team that then knows the screws for targeted project communication, but also the managers and employee representatives—as multipliers—can point out different perspectives to the other colleagues in a personal exchange.

In communication with employees, however, it is not enough to convey only rational information. Our willingness to change is significantly influenced by information we receive at the emotional level. For example, employees become more involved in the forthcoming change if they perceive a positive climate of change and can follow the company's vision (ibid., p. 122). The willingness to embrace the new is thus influenced by the prevailing mood in the company. The key factor here is leadership. Employees must feel supported both by the company management and by their own managers (ibid.). Even if companies think they support their employees sufficiently, they should self-critically check how the current form of support is received. In practice, employees feel highly valued when representatives of the company management also take part in the above-mentioned coffee talks or regulars' table events. In another practical example, information events were held to present the new workplace concept at departmental level. In addition to the project management, which presented the rational information, a board member of the company also took part in numerous events. Every employee had the opportunity to talk directly to the company management. He reaffirmed the company's vision and decisions in the event of queries, asked for new ideas, openly took on board the fears and thoughts of the employees and referred to the design options available to each department. All hierarchical levels took part in the event at the same time. Corporate management, management level one, management levels two and three as well as the employees. All received the information at the same time with the request to enter into dialogue together in the coming weeks. For this purpose, there was a set date limit for determining the scope for design. Information losses caused by a cascading of the information flow from "top" to "bottom" were encountered in this way and a common sense of optimism was created in every department.

Managers as the Key to Successful Change

The intensive accompaniment of the introduction of a new working environment through a change management programme has proven to be necessary and proven in many cases at the Fraunhofer IAO. The change process must be professionally managed and accompanied in order to successfully realise the vision of the company and the associated goals. The following applies: The more comprehensive the change, the more broadly the implementation of the vision must take place and the more broadly change measures should reach the employees.

Each organisational unit should work out for itself what opportunities the future working environment of the unit offers and how the opportunities can be realised jointly—as a unit. Middle management executives have a crucial role to play in this context. They are caught between the fears and uncertainties of the workforce and the strategic and operational requirements of senior management (Rouleau 2005). As key persons in the change process, this level needs differentiated development offers. Specific seminars on leadership in change processes in general and in the concrete change process are strongly recommended based on practical experience. In the context of new working environments, managers are often faced with a change in spatial structures. In addition to functional and organisational challenges, this can pose a threat to one's own status (cf. Fischer 1990). Work and management processes must be reflected upon and reorganised. The willingness to implement should exist in order to support the employees in the change process. If managers do not accept a change in their own way of working, how can they successfully guide their employees through the change process?

The vision must be carried down to the lowest management levels in order to be able to meet questions, fears, resistance, etc. on the part of the employees regularly in dialogue. Once again, strong leadership at all levels is required instead of pure management. Recognised personalities exert a strong force on the workforce, both positively and negatively. It is therefore highly recommended to identify such recognised personalities and to actively involve them in the project by taking over sub-projects. This gives them the opportunity to design project themes and is a visible supporter of the change process. Any company planning to introduce a new world of work must understand that its leaders are the role model for the desired change (Melkonian 2005; Simons 2002)! A manager should therefore embody, as consistently as possible, that he or she will consistently lead the achievement of the vision of the future. Contradictions between the actions of the manager and the vision of the future can lead to doubts and demotivation among the employees and thus reinforce a rejection of the future office environment.

Example

An example from practice is described as follows: A large company integrated relaxation rooms into its new office concept. These were of high quality and very invitingly designed. However, some time after the new working environment was put into operation, it turned out that it was not being used. As outlined at the beginning of this chapter, retreats for recovery have a significant impact on health (90% of respondents to a study said they had a need for retreats to switch off during working hours; Wackernagel 2017). Regeneration of performance during working hours is essential, as a permanently maintained stress level can cause serious damage to health and well-being (Geurts et al. 2005). In practice, managers regularly take the view that work (presence in the company) is done exclusively at work and that regeneration takes place exclusively in leisure time. In our practical example, the employees assumed that their managers would take this position, and nobody wanted to be the one "lying around lazily". After the managers were sensitised to this topic and able to reflect on their own behaviour, they took on their role as role models. The use of the relaxation rooms by the managers meant that the rooms were accepted by the employees. Accompanying change measures for managers thus directly influences the acceptance of the new working environment on the part of the employees. However, the perceived support of employees by managers is most effectively promoted through regular individual discussions and exchange opportunities (cf. Schott and Jöns 2004). That is in the context of a new world of work en are also the case.

Involving Employees in the Change Process

Organisational change can be experienced by employees as being difficult to control. The experienced anxiety then leads to negative stress experience. A perceived controllability within the change process, on the other hand, increases the willingness to perform and reduces the feeling of stress. (cf. Brotheridge 2003). Project-related activities increase the perceived control of employees (cf. Larsson and Lubatkin 2001) and create a sense of optimism in the workforce. For this reason, a targeted involvement of employees or employee representatives is indispensable. In addition to managers, employees must also be given the opportunity to deal in depth with the future world of work. There is a wide range of measures that can be implemented as part of a change management programme. In addition to the above-mentioned possibilities of preparing employees for the change to the new working world, only essential participation possibilities are described below.

A significant driver of change at the employee level is so-called **change agents** (Huy 2002). Change agents are employee representatives who are used as multipliers in the company. In practice, change agents are contact persons for their colleagues when it comes to content and organisational questions about the project and represent their colleagues in shaping the future working environment. At least one employee per organisational unit should represent the department as a change agent. A critical success factor in practice is the choice of employees willing to communicate for this role, the skills for good communication can subsequently be developed through training measures. Change agents are proactive contact persons and are intended to encourage colleagues to exchange ideas. In addition, they should convey confidence in success and feel able to accept critical questions from colleagues. The inclusion of the role should, however, take place exclusively on a voluntary basis. In practice, it has been shown that the appointment of employee representatives can significantly impair the change process. Furthermore, a balanced relationship between the employees to be supported and the change agents should be strived for. Here, too, a direct dialogue between colleagues should be possible, which is often difficult to achieve in practice with a number of more than twenty employees per change agent. In order to win employees for this role in the change process and to reduce stress, capacities must be made available to the change agents. In practice, this value is between 10 and 20% of the workload. This timeframe is needed for participation in working sessions, sub-projects to design and implement the new office concept, training, communication with colleagues in the department, etc. Companies must bear in mind that change agents are responsible for the change process. both as affected persons and as supporters of change. As a result, they are exposed to strong pressure (Piderit 2000). Regular reflection of the mood among the change agents is essential in order to be able to provide assistance in dealing with internal and external resistance.

In order to involve as many employees as possible in shaping the future working environment, **employee surveys** are frequently used in practice. Employees are given the opportunity to communicate their satisfaction with the working environment via a before-and-after survey. The experience of available from many employees surveys. The study on the introduction of a modern workplace concept (combined with the introduction of an open office structure) shows that, in principle, a significant increase in satisfaction is achieved. The first survey must be carried out as early as possible in the project in order to avoid disruptive effects (reactions to certain topics in the change process). Here, the current working environment is analysed and its weak points are determined. By optimising the identified adjusting screws, employees can be won over to the new world of work. The follow-up survey takes place at the earliest three months after moving into the new working environment, when the employees have become accustomed and the new office concept has become familiar. A further survey at a third point in time (8–12 months after moving in) can in principle be recommended, as the satisfaction with the increasing acclimatisation increases further. But there may also be a need for optimisation in the new office concept. In practice, many companies are reluctant to measure their satisfaction with the working environment. After all, what do companies do when the results are worse? It is important to take the information provided by employees seriously and to further promote employee satisfaction and acceptance through structural, design, behavioural or knowledge-enhancing measures.

In practice, employees are better prepared for a change if the design of an office concept is not rigidly fixed, but remains adaptable even after the move-in. This makes employees more willing to try something out and get involved in something new. The principle of trial and error is also behind the establishment of a **pilot area**. A test area allows the company and its staff to try out the new office concept before it is rolled out throughout the company. For this purpose, one or more departments are selected which gain experience in the test area for a defined period of time. For the user, the future working environment becomes tangible, tangible and tangible in the real sense. Even if only part of the workforce is working there, the other employees can visit the pilot area (at regular times) and discuss the advantages and disadvantages with their colleagues. A pilot area should in principle be accompanied by a before-and-after evaluation in order to optimise it with the experiences of the pilot users before it is introduced into the company as a whole.

The development of a **code of conduct** is indispensable for the introduction of an open office structure. The open working environment requires a different way of use than, for example, a cell structure. All employees must be made aware of this. The code of conduct can be developed by different user groups and with different scopes. There are companies that interpret the rules (see section 'Dealing with Colleagues' in Chap. 10). In other companies, the code of conduct is drawn up with employee representatives on each floor. It is recommended that a code of conduct should always be drawn up jointly by user groups that also use common infrastructures. A study (Jahncke 2017) examined the ability to concentrate in different rooms of a modern office concept. The researcher specifically examined the concentration performance in an open area in which activities take place without restrictions and no code of conduct exists, as well as an open area which is defined as a quiet area and in which a code of conduct therefore applies. In addition, the concentration performance in a project room, meeting room, retreat room with individual workstation as well as in the lounge area was examined. In retreat areas, in which the user was alone, the concentration performance was comprehensibly highest. In the open area with a code of conduct, the concentration performance is lower, but much higher, than in the open area without a code of conduct. The code of conduct is an explicit agreement that must be supported by all employees. In practice, there are always voices that say "We are all adults, something ridiculous like that" or "We all have common sense". social psychology The results of the Hielscher studies (Hielscher 2014) show that we humans unconsciously adapt to the apparently valid norm. Experience from numerous workshops to develop a code of conduct has shown that even one or two standing coffee cups in the kitchen can cause users to place the cup in front of the others rather than check to see whether the dishwasher is actually empty or to clear it out. It requires a mental effort from us to overcome automatisms. In order to achieve a high level of acceptance and thus compliance with the code of conduct, it is essential to develop it on a company- or division-specific basis.

Creating Continuous Profits for Employees

Change projects usually have a long duration. In practice, it takes an average of two to three years from the decision to introduce a new working environment to the move-in—and thus the implementation of the change. Companies should offer attractive participation opportunities for employees throughout the entire process in order to maintain the spirit of optimism and to convey the fun of change. Info days, in-house exhibitions, topping-out ceremonies, construction site visits and guided tours through reference examples satisfy the curiosity of employees and are only a few examples of interesting events on the way to the future world of work. Some topics of the new working environment can be introduced from the fields of technology/IT and personnel development before moving into the new working environment in order to encourage a more flexible and self-determined way of working at an early stage.

According to a study by the Fraunhofer IAO (Rief et al. 2014), the free choice of means and methods to achieve work goals, temporal autonomy and the choice of the place of work have a positive effect on many areas of work and private life. Employees who can design their work individually experience a higher work–life balance, more motivation and more performance. In conversations with employees and managers, the strong desire for a **home office** is always expressed (or third-place working). This does not mean that employees no longer want to come to the office. You wish to be able to decide flexibly to work on work tasks from an external location (at home, while travelling, at partner companies, etc.). Not every company already has a uniform home office regulation for all employees. In a current trend analysis conducted by ISF, Munich, approximately 60% of the companies surveyed offered managers and employees the opportunity to work in their home office. Even if the possibility of working from home is not an explicit measure of change management, home office can still support the employees' ability to change and represent a strong profit for the employees within the change process.

Anchoring the Change in the Corporate Culture

It is essential that the vision developed and the goals set by the company management are increasingly and extensively anchored in the corporate culture. This makes it clear to employees and managers that the new world of work is a sustainable and irreversible process. For us, environments are symbol bearers (Saegert and Winkel 1990). Every single part of a building is a sign that conveys a message to the user. By redesigning an office environment, companies have the opportunity to decisively determine which contents of their own corporate culture should be communicated to their employees: Joy at work, cost pressure appreciation , steep hierarchy n, creativity directive leadership, cooperation, work by the book, liveliness, secrecy, exchange of knowledge, quality work... a list would be infinite. Companies should use the symbol of the working environment to make the future vision tangible for the user in their daily work.

However, when organisations have gone through a lot of changes in a short period of time, certain fatigue of change can occur among a broad workforce. It is not uncommon for several drastic changes within an organisation to occur in parallel. In a concrete case study, the employees were confronted with five further change projects in addition to the introduction of a new working environment! Only the project to introduce the new working environment was accompanied by change management measures. Parallel changes can offer enormous synergy effects for an organisation. Technologies, processes, organisational structures and working environment must be coordinated with each other and with the desired corporate culture. Nevertheless, as mentioned above, changes can lead to anxiety and stress. If these occur on a broad basis, individual stress reactions of the individual concerned can be transferred to the organisation. With the stress experienced collectively, the symptoms also occur collectively (Glazinski 2004). In this way, change processes are blocked or inhibited if necessary, without the individual change being rejected per se. Extensive changes require a holistic change management programme that is well tuned to the individual components and supports the entire workforce in the long term.

The Health-Promoting Use of an Office Environment

In this chapter, companies were shown many ways to be considerate of the health of their employees when developing a new working environment. Managers of a study (Udris et al. 1994, p. 200) answered the question *Who in the company is responsible for the health of employees?* with the basic statement: *Primarily, employees are responsible for their own health.* This makes it clear that health is often not understood as a management task. Due to the duty of care, however, the protection of health is to be understood as an original management task (Rudow 2014, p. 320). How can companies and managers support a health-promoting behaviour of their employees—in the sense of self-care? Once a health-promoting working environment has been created, employees should be taught which forms of use are good for their own health and which can damage it. A one-off clarification on the subject will cause only little correction in the accustomed and possibly damaging behaviour of employees. Not only must knowledge be enriched, but companies must achieve the health-related attitudes of their employees (ibid., p. 325). A working environment should therefore be designed in such a way that it involuntarily prompts the user to move. Past office environments were designed to be efficient but not conducive to movement. They created a bond to the workplace by ergonomically optimising the gripping space and the arrangement of all work equipment according to the short-path principle. The result we see today is that 30% of all illnesses are related to the musculoskeletal system (BAuA 2010). In recent years, researchers have provided alarming figures (Wallmann-Sperlich et al. 2014; WHO 2010; Chomistek et al. 2013; Hamilton et al. 2007; Owen et al. 2009). We spend 80% of our working time sitting at our desk. In addition, 42% of working people work less than 0.5 h a day. In comparison with a *low seater n* (three hours a day), the risk increases in the case of *much seater n* (six hours a day) to die in the next 15 years by 40%. We should be alarmed because our average sitting time on a weekday is 7.5 h. In addition, there is the bad news that sitting for hours on end is an independent risk factor—even daily sport cannot compensate for the consequences. Standing calls in the working environment such as standing meeting possibilities, standing desks and high tables with standing aids signal position changes to the user and should be invitingly designed. The use of height-adjustable tables, on the other hand, must be trained. Since the metabolism completely shuts down after two hours of sitting (consumption: one calorie per minute), users should be encouraged to actively adjust the height of the table. One person makes every phone call standing up, the other raises the table at regular intervals. Rituals enable us not to have to consciously think about a change of position in our daily work. The targeted "banning" of frequently used utensils (files, working materials, printers, etc.) from the gripping area is a design measure that promotes movement, even if users frequently complain about it in practice. Of course, this must not result in long journey times, but frequent and short changes of posture maintain our metabolism. A high-quality range of different space offers (retreat spaces, project space, meeting rooms, informal meeting zones, creative area meet the specific requirements of different work processes. The choice of the room according to the activity automatically results in a short interruption of the sitting activity. Employees should find a wide range of seating and standing options in the additional rooms and areas. In contrast to the standard workstations, in the additional rooms varied seating is more important than fully fledged, highly ergonomic office chairs. The length of stay in the additional rooms is predominantly limited, because the different room offers should be available for each employee—independently of the hierarchy level—to suit the respective activity. In this way, every user can enjoy the same advantages of the respective room module. Breaks from work are also important for our health. Shorter work breaks have a stronger recreational effect on people than a few long work breaks of the same total duration (BAuA 2010). Lounge areas and generously designed coffee kitchens close to the workplace support short breaks from work. In order to sustainably strengthen all these health-promoting behaviours among employees, managers themselves must exemplify the new use of the office environment.

Bottom Line

In order to protect and promote the health of employees, there are two basic approaches: the creation of health-promoting conditions and the promotion of healthy behaviour. In the context of new working environments, the design of a health-promoting office environment is the basis for protecting employees. Since satisfaction with the office environment is significantly influenced by individual or collective perception and evaluation, the requirements of the employees should be incorporated into the future office concept. The human being is influenced in his attitude formation by what is known and familiar to him. In order to support the employees' willingness to change, they need extensive information and opportunities for participation in the project. The employee must understand why a new office structure is a win–win situation for the company, the workforce and for himself. A powerful vision of the company management and good leadership through the change process also show the workforce comprehensible and attractive goals that can only be achieved in the community. As drivers for change, managers need differentiated development opportunities, because all management levels take on a role model function in the change process, which employees use as a guide. Employee representatives (change agents) are used as multipliers of project information and are also involved in sub-projects for the design of the future office concept. In addition, change management measures adapted to the company and its employees accompany the workforce through the entire change process until the new working environment is familiar and the new ways of use have been introduced. In order to promote healthy use, the new world of work should be designed in such a way that the user is involuntarily encouraged to exercise. Training courses and the role model function of managers also contribute to ensuring that health-promoting behaviours are lived out sustainably by employees. Ultimately, organisations can significantly develop their work culture through a well-designed change process and, together with their employees, create a healthy working environment of tomorrow.

1.2 Work–Life Integration—Harmonious Working Environments for the Well-Being of the Employees

Working environments change constantly and in ever faster cycles. So speed, mobility, adjustment and versatility will be the new success criteria for the office of the future. But if you can decide for yourself in future where, when and how you want to work, will you then still need offices at all? This is to be answered quite clearly with "yes", because it lies in us humans that we want to belong to a "cool" community and would like to feel belonging and well. This requires the appropriate rooms and the right furniture in a harmonious office environment with the right meeting and retreat areas.

Today, a good workplace means that employees feel at home and that as an organisation, as an employer, one cares for one's employees and that they are and remain healthy and productive in order to develop creative and innovative solutions and products. But what does "feeling good" mean to people? Feeling pleasant and comfortable means having a good mood and being healthy. As defined by the World Health Organization, health includes mental, social and physical health. Feeling good means feeling happy, having inner peace and being well balanced in the daily hustle and bustle and the different forms of stress at work. Feeling very good also means developing initiative for one's own health and satisfaction. The subjective feeling of well-being in psychology is the feeling of happiness in life and satisfaction with one's own life. All in all, this simply means quality of life.

If you consider the social and mental well-being at work, this means that you feel important, you know that you and your own work are recognised and that you can make an important contribution to the work of the company through your own knowledge. For any organisation, the well-being of its employees is crucial to be creative and to achieve the critical innovations to be successful.

Studies show that health impairments can occur if the number of employees in an office increases (Lucerne University of Applied Sciences and Arts, Seco 2010). Illness-related absenteeism is higher in multi-person offices than in individual offices. But in order to achieve a high level of cooperation—very important for knowledge societies—, we cannot retreat into individual offices, but we need transparency and spaces for exchange and encounter. In order to avoid health risks, it is important to plan office space well and sufficiently and to consider all aspects that are important for well-being in the workplace. In addition to good acoustics, light and climate, this also includes taking employees along and accompanying them right from the start. A good change process supports understanding, addresses all opportunities and challenges and accompanies you through all phases of the project, including difficult ones.

Terms such as work–life balance are becoming more important again and are mutating more into a work–life integration. You think about really important values in life and people who are important to you and with whom you want to spend time. Young family fathers and mothers want to see their children grow up and experience developmental progress. For example, employees would like to decide for themselves whether they want to attend kindergarten or school performances in the afternoon and work through the unfinished work at home in the evening when the children are in bed. Older workers may have to take care of their own parents and also need more flexibility in time management and the freedom to choose where to do individual tasks. For some, home office might be a good solution; others prefer to separate work and family and therefore do not want to work at home. Here, too, flexibility is required and the involvement of the employees is required in order on the one hand to find a tailor-made solution and on the other hand to decide on the right additional offers for the organisation (Fig. 1.2).

This could be care services for children a sports room, like gymnastics or fitness room; quiet recreation rooms, like a library for quiet relaxation; or rather louder rooms with a soccer table, billiards or pinball machines. Employees may also want a post office or a concierge service where they can order theatre or other event tickets. A small gift shop or a grocery store, for example, is also conceivable. The offers

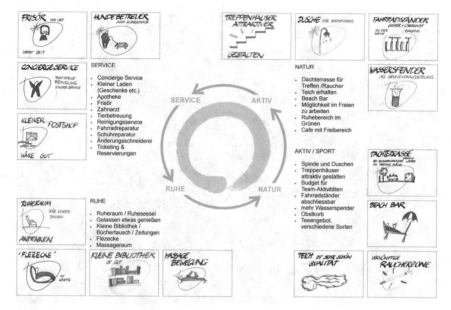

Fig. 1.2 Example of services for employees—offers one could imagine

here are very diverse and include group activities such as yoga courses, up to professional individual consulting and personally tailored training plans. It is important that employees have a certain say that they can participate in the selection process and that later there is something like a common consensus about the selected options and that there are also sufficient health-relevant offers. There does not necessarily have to be a fitness room directly in the company, a cooperation with a company or a fitness club nearby can also be offered. The possibility of borrowing e-bikes also promotes daily exercise and stimulates some people to give up their car from time to time.

The spatial organisation of the layout can also contribute to the health of the employees. Beautiful stairwells invite you to use them, and if they are well arranged and wide enough, they connect departments and different floors well with each other and thus also invite you to stop and exchange ideas with colleagues (Fig. 1.3).

This also applies for tea kitchens, which no longer should be hidden in leftover dark, windowless residual spaces, but deserve attention and thus become a multi-functional space. In a beautifully designed and inviting tea kitchen, one likes to meet, to welcome visitors, the room also can be used for meetings and in quiet periods for quiet work, like reading or a for making private telephone call (Figs. 1.4 and 1.5). In this context, it should also be pointed out that the sanitary areas are well designed. This promotes hygiene and shows the employee that he or she is important and valued.

The promise of flexibility and own time management lures employees into the job, but that alone is not enough to keep them there. Young creative employees need com-

Fig. 1.3 Example of inviting staircases (University of Kolding Henning Larsen Architects, Photo: Hufton Crow)

Fig. 1.4 Multifunctional tea kitchen (Drees & Sommer Photo: Christine Kohlert, Stuttgart)

Fig. 1.5 Kitchenette can also be used for individual work (RBSGROUP, Munich Photo Peter Neusser)

mitment and engagement, interesting and surprising spaces as well as experiences and an inspiring environment. In order to emotionally bind employees to the company, it is of great importance to make one's own brand values tangible and to transport them internally through an authentic design of the working world. Workspaces that convey basic human needs such as security, recognition and self-realisation are emotionally binding and enable free thinking and the necessary attention.

A good mood makes people more tolerable, healthier and encourages them to think more creatively. An optimistic and authentic environment that gives employees choices and control, and supports movement and interaction, also contributes to this. In addition to these spatial aspects, it is also about meaningful activities and mutual attentiveness and about providing employees with places for exchange and cooperation as well as for peace and concentration.

Social exchange and maintaining contacts are important and can be promoted and supported by a good spatial environment. A good culture of trust also contributes to this, which is above all about achieving goals and not about a culture of presence. Only in such a good relationship of trust and cooperation can a company be truly successful and achieve top performance. Motivated employees bring themselves and their knowledge to the table and work out the right solutions together. But such a culture must always be exemplified by the superior and he/she must convey to everyone that the entire office with all its facilities is a place for everyone, a marketplace of knowledge and a place for exchange and common thinking and working. This is the only way to work together with colleagues and create a good relationship between employee and employer for the benefit of all. At the end of the day, the working atmosphere is passed on to the customer and thus the esteem that contributes to success.

Ultimately, management determines how much trust prevails in the company and how open corporate communication is. Instead of rigid, hierarchical decision-making structures, ideas, customer orientation and speed are in demand in the new business world. Creativity and new ideas only emerge when discussion is allowed and everyone's opinion counts; lengthy decision-making processes inhibit idea generation and dynamics. New forms of work and organisation are needed, with more teams, shorter distances, fewer hierarchies and more cooperation, also across departments and with impulses from outside.

Bosses become coaches who listen, praise, encourage and willingly give feedback. They have a clear vision and develop strategies to inspire themselves and others. They encourage and support their employees, transfer responsibility to them and give them the freedom they need to do so. They are straightforward and just have a feeling for their own strength and enough self-reflection to be able to lead in a humane way. They see their organisation as a functioning family that is treated with care and trust. Thus, a natural code of conduct develops, which can be further positively supported with rules of behaviour, which are agreed upon together.

In addition to all these aspects, sustainability and sharing in a sharing economy are becoming increasingly important today, and biophilic design also contributes to further well-being. Plants can reduce stress and make people happier, improve air quality and enhance productivity and memory (Fig. 1.6; Sect. 10.1 Which are the suitable plants?).

Light, especially sunlight, improves the quality of life and ensures healthy sleep, reduces stress and anxiety and increases productivity. View to the outside relaxes the tired eyes from the constant fixation on the screen. Sustainability goes hand in hand with the desire for "real" materials that create a "haptic experience" and feel good. Buildings and rooms thus also convey a "feeling". Nature is not rectangular, so curves and natural shapes also stand for well-being. A harmonious design is rounded off by a harmonious and balanced choice of colours. Colour modifies the light and determines the atmosphere; it distinguishes the various elements in architecture and directs our attention to the most important and it enables contrast and individuality. Colour must always be seen in context, the right combination creates a positive visual experience (Fig. 1.7).

Qualifications for the knowledge workers of the future are fast adaptability, creative thinking and knowledge into the right context. A harmonious workplace can support them and contribute to satisfaction and well-being. Planning should be based on six principles.

Principle 1: There are no guarantees, but spaces are needed for different activities such as concentration and communication. These rooms support creativity and learning and allow employees to make their own choices (Fig. 1.8).
Principle 2: Comfort is key. All dimensions of well-being must be met, as well as physical comfort, starting from the workplace to common rooms, retreats and recreation areas (Fig. 1.9).
Principle 3: Space can release good behaviour and makes us aware of who and what is present. Naturally exposed places for communication and exchange with a view the countryside promotes this attitude.
Principle 4: Flexibility and variability are an absolute necessity in order to be able to react quickly and efficiently to changing conditions (Fig. 1.10).
Principle 5: Space in connection with nature is desirable, i.e. bringing green as much as possible to the inside and making views possible.
Principle 6: A room is only as good as those that lead in it. Ultimately, management determines how many feel good factors are implemented and how they are used—mutual trust is the most important thing.

All in all, it is a matter of perfectly transforming the relationships that are important for achieving the entrepreneurial goals into space. Good rooms are always a contribution to the company value. Like in a good restaurant the overall harmony counts, from the welcome to the farewell of the guest. the entire office with its environment and organisation makes a decisive contribution to a harmonious working environment that leads to well-being and satisfaction of the employees and thus ensures the success of a company.

(a)

(b)

Fig. 1.6 Green and natural materials enhance the mood and contribute to well-being and a good climate (photo left): Google Zurich, interior design Evolution Design, photo: Peter Wurmli; right: Rapt Studio, San Francisco, interior design O+A) (left photo: Peter Wurmli, lines: Christine Kohlert)

Fig. 1.7 Relax in rooms flooded with light and air (Microsoft Lyngby, Denmark) (Architecture Henning Larsen, Photo: Hufton Crow)

Fig. 1.8 Various office options—Rooms for concentration and communication (Booking com Salzburg, Photo Peter Neusser)

Fig. 1.9 Comfortable furniture throughout the company (Easy Credit, Nuremberg) (Interior design Evolution Design, photo Christian Beutler)

Fig. 1.10 Rooms for communication and exchange and all kinds of cooperation (interior design studio O+A, photo Jasper Sanidad)

Literature

Altman, I. (1970). Territorial behavior in humans: An analysis of the concepts. In L. A. Pastalan & D. H. Carson (Eds.), *Spatial behavior of older people*. Ann Arbor.

BAuA. (2010): *Well-being in the office: Occupational health and safety in office work*. Available under Federal Institute for Occupational Safety and Health. http://www.baua.de/de/Publikationen/Broschueren/A11.html. Accessed on 5 October 2017.

Bodin Danielsson, C. (2013). An explorative review of the lean office concept. *Journal of Corporate Real Estate Planning Research, 15*(3–4), 167–180.

Bodin Danielsson, C., & Bodin, L. (2008). Office type in relation to health, well-being, and job satisfaction among employees. *Environment and Behavior, 40*(5), 636–668.

Brotheridge, C. M., & Lee, R. T. (2003). Development and validation of the emotional labor scale. *Journal of Occupational and Organizational Psychology, 76*, 365–379.

Canter, D., & Rees, K. (1982). A multivariate model of housing satisfaction. *International Review of Applied Psychology, 31*, 185–208.

Chomistek, A. K., Manson, J. E., Stefanick, M. L., Lu, B., Sands-Lincoln, M., Going, S. B., et al. (2013). Relationship of sedentary behavior and physical activity to incident cardiovascular disease: Results from the women's health initiative. *Journal of the American College of Cardiology, 61*, 2346–2354.

De Croon, E. M., Sluiter, J. K., Kuijer, P. P. F. M., & Frings-Dresen, M. H. W. (2005). The effect of office concepts on worker health and performance: A systematic review of the literature. *Ergonomics, 48*(2), 119–134.

Dijksterhuis, A., Bos, M. W., Nordgren, L. F., & van Baaren, R. B. (2006). On making the right choice: The deliberation-without-attention effect. *Science, 311*, 1005–1007.

Doppler, K., & Lauterburg, C. (2014). Change management. shaping corporate change (13th ed.) Frankfurt: AUFL/M., New York: Campus Verlag.

Fischer, G. N. (1990). *Psychology of the workroom*. Frankfurt/M.: Campus Verlag.

Geurts, S. A. E., Taris, T. W., Kompier, M. A. J., Dikkers, J. S. E., van Hooff, M. L. M., & Kinnunen, U. M. (2005). Work-home interaction from a work psychological perspective: Development and validation of a new questionnaire, the SWING. *Work & Stress, 19*(4), 319–339.

Glazinski, B. (2004). *Strategic corporate development: identifying and averting crisis signals at an early stage*. Wiesbaden: Gabler Publishers.

Hamilton, M. T., Hamilton, D. G., & Zderic, T. W. (2007). The role of low energy expenditure and sitting on obesity, metabolic syndrome, type 2 diabetes, and cardiovascular disease. *Diabetes, 56*, 2655–2667.

Hertel, G., van der Heijden, B., De Lange, A., & Deller, J. (guest Eds.). (2013). Facilitating age diversity in organisations—Part I: Challenging popular misbeliefs. Special issue of the *Journal of Managerial Psychology, 28*, 729–856.

Hielscher, D. R. (2014). *The broken windows theory: On the influence of the spatial environment on social behaviour*. Berlin: epubli.

Huy, Q. (2002). Emotional balancing: The role of middle managers in radical change. *Administrative Science Quarterly, 47*, 31–69.

Lucerne University of Applied Sciences and Arts, Seco. (2010). SBiB study: Swiss survey in offices. https://www.news.admin.ch/NSBSubscriber/message/attachments/18922.pdf. Accessed on 5 October 2017.

Jahncke, H. (2017, August). *Activity-based workplaces: Changes in cognitive performance among worker previously employed at cellular offices or open-plan offices*. Article Presented at the International Conference of Environmental Psychology: Theories of Change and Social Innovation in Transitions Towards Sustainability, A Coruña, Spain.

Kavanagh, M. H., & Ashkanasy, N. M. (2006). The impact of leadership and change management strategy on organizational culture and individual acceptance of change during a merger. *British Journal of Management, 17*(1), 81–103.

Klinger, C., Curth, S., Müller, C., & Nerdinger, F. W. (2014). *Older employees in the innovation process. An explorative interview study. Rostocker Beiträge zur Wirtschafts- und Organisationspsychologie, No. 14*. Rostock: University of Rostock, Chair of Economic and Organizational Psychology.

Kotter, J. (2012). *Leading change*. Boston: Harvard Business School Press.

Lauer, T. (2014). *Change management: fundamentals and success factors* (2nd ed.). Berlin, Heidelberg: Springer Verlag.

Landes, M., & Steiner, E. (2014). *Psychological effects of change processes. Resistance, emotions, willingness to change and implications for managers*. Wiesbaden: Springer VS.

Larsson, R., & Lubatkin, M. (2001). Achieving acculturation in mergers and acquisitions: An international case study. *Human Relations, 54*(12), 1573–1607.

Lazarus, R. S., & Folkman, S. (1984). *Stress, appraisal, and coping*. New York: Springer.

Lütke Lanfer, S., & Pauls, N. (2017, February). *A systematic literature search on modern office structures and their impact on mental well-being*. Article presented at the 63rd Spring Congress 2017 of the Gesellschaft für Arbeitswissenschaft e. V.: Socio-technical design of digital change—creative, innovative, meaningful, Brugg and Zurich, Switzerland.

Melkonian, T. (2005). Top executives reactions to change: the role of justice and exemplarity. *International Studies of Management and Organization, 34*(4), 7–28.

McFarlin, D. B., & Sweeney, P. D. (1992). Research notes. distributive and procedural justice as predictors of satisfaction with personal and organizational outcomes. *Academy of Management Journal, 33/3*, 626–637.

Nasar, J. L. (1994). Urban design aesthetics: The evaluative quality of building exteriors. *Environment and Behavior, 26*, 377–401.

Oommen, V. G., Knowles, M., & Zhao, I. (2008). Should health service managers embrace open plan work environments? A review. *Asia Pacific Journal of Health Management, 3*(2), 37–43.

Owen, N., Bauman, A., & Brown, W. (2009). Too much sitting: A novel and important predictor of chronic disease risk? *British Journal of Sports Medicine, 43*(2), 81–83.

Piderit, S. K. (2000). Rethinking resistance and recognizing ambivalence: a multidimensional view of attitudes toward an organizational change. *Academy of Management Review, 25*, 783–794.

Rief, S. (2017, March). *The effect of space on motivation, performance and well-being at work*. Article presented at the 39th Uponor Congress 2017, St. Christoph am Arlberg, Austria.

Rief, S., Jurecic, M., Kelter, J., & Stolze, D. (2014). Short report on the study office settings. https://www.office21.de/content/dam/office21/de/documents/Publikationen/Fraunhofer-IAO_Office-Settings.pdf_short_report. Accessed on 5 October 2017.

Rouleau, L. (2005). Micro-practices of strategic sensemaking and sensegiving: How middle managers interpret and sell change every day. *Journal of Management Studies, 42*, 1413–1443.

Rudow, B. (2014). *Healthy work. Psychological stress, work design and work organisation*. Berlin/Munich: De Gruyter Oldenbourg Science Publishers.

Simons, T. L. (2002). Behavioral integrity: The perceived alignment between managers' words and deeds as a research focus. *Organization Science, 13*, 18–35.

Slesina, W. (2001). Evaluation of occupational health circles. In H. Pfaff & W. Slesina (Eds.), *Effective workplace health promotion* (pp. 75–95). Weinheim: Juventa.

Szebel, A. (2015). *Change competence of employees. An empirical study on the differential development of the construct of the individual change competence of employees with special consideration of the influence of dispositional personality factors*. Dissertation, University of Cologne.

Saegert, S., & Winkel, G. H. (1990). Environmental psychology. *Annual Review of Psychology, 41*, 441–477.

Samuelson, W., & Zeckhauser, R. (1988). Status quo bias in decision making. *Journal of Risk and Uncertainty, 1*, 7–59.

Schott, U., & Jöns, I. (2004). Change of attitude in mergers: An integrative model for the effect of information and communication. *Mannheimer Beiträge zur Wirtschafts- und Organisationspsychologie, 2*, 53–59.

Schweiger, D., & DeNisi, A. (1991). Communication with employees following a merger: A longitudinal experiment. *Academy of Management Journal, 34*(1), 110–135.

Udris, I., Rimann, M., & Thalmann, K. (1994). Maintaining health, producing health: On the function of salutogenetic resources. In B. Bergmann & P. Richter (Ed.), *The Theory of Action Regulation. From the practice of a theory* (pp. 198–215). Göttingen: Hogrefe.

Ulich, E., & Wülser, M. (2015). Health management in companies. In *Work psychological perspectives. Six.* Wiesbaden: Springer Gabler.

Wackernagel, S. (2017, February). *How the design of the office work environment influences our well-being.* Article presented at the 63rd Spring Congress 2017 of the Gesellschaft für Arbeitswissenschaft e. V.: Socio-technical design of digital change—Creative, innovative, meaningful. Brugg and Zurich, Switzerland.

Wallmann-Sperlich, B., Bucksch, J., Schneider, S., & Froböse, I. (2014). Sitting as a risk factor: Prevalence and determinants of sitting times at the workplace. *Healthcare, 76*(8–9), A209.

WHO. (2010). *Global Recommendations on Physical Activity for Health.* Geneva: WHO Press.

Windlinger, L. (2017, May). *Office quality: Construction, operation and users' view.* Article presented at the 2nd Symposium of the Fraunhofer Institute for Building Physics: People in Spaces, Stuttgart, Germany.

Windlinger, L., & Zäch, N. (2007). Perception of stress and well-being in different office forms. *Zeitschrift für Arbeitswissenschaft, 61*(2), 77–85.

Chapter 2
Perceptions

Werner Seiferlein, Ulrich Pfeiffenberger, Gerd Danner, Torsten Braun
and Christine Kohlert

2.1 Introduction to the Chapter: "Perception"

What does perception mean?

Perception is what you feel or notice with your senses. These are stimuli that influence the outside world (cf. Spath et al. 2011, p. 30). Five senses are defined:

- Odour,
- Vision,
- Hearing,
- Tastes,
- Feel.

Feeling or the sense of touch can be integrated again in the units

- Contact
- Pain and temperature (surface sensitivity),

W. Seiferlein (✉)
Technology Innovation Management, Frankfurt/Main, Germany
e-mail: werner.seiferlein@timoffice.de

U. Pfeiffenberger
Engineering Company Pfeiffenberger mbH, Neu-Isenburg, Germany
e-mail: u.pfeiffenberger@t-online.de

G. Danner
SoundComfort GmbH, Berlin, Germany
e-mail: g.danner@soundcomfort.de

T. Braun
The Light Planners, Limburg Relay, Germany
e-mail: limburg@lichtplaner.com

C. Kohlert
RBSGROUP Part of Drees & Summer, Munich, Germany
e-mail: christine.kohlert@rbsgroup.eu

© Springer Nature Switzerland AG 2020
W. Seiferlein and C. Kohlert (eds.), *The Networked Health-Relevant Factors for Office Buildings*, https://doi.org/10.1007/978-3-030-22022-8_2

but also active recognition like the

• Haptic perception

Can be subdivided. Haptics is a part of building designs and its interiors (cf. Chap. 4 adequate office equipment). The combination of materials is becoming more and more common and more consciously.

More and more companies are using tactile perception to make their products work with senses that are not so often used: Flensburger has brought the beer bottle with the Plopp cap back onto the market and advertised it—the customer can touch the bottle, see, feel and hear how the bottle opens (Plopp) (cf. Schmitz 2014).

The influencing factors illuminated here occur as mutual perceptions as so-called $4 \times Ls$ (©Seiferlein $4 \times Ls$). In the field of real estate, real estate buying and selling, one would then refer to the synonym $3 \times Ls$ as 'location, location, location'. Here, there are four factors that are described as follows: **air, Light, Noise and body**.

These four factors are directly related to human perception.

Why are we talking about the $4 \times Ls$ in this book?

The factors air and light bind and integrate the impression and idea of sitting in fresh air and daylight. This nourishes the feeling of sitting outside.

When people complaints, then very often to the factors air, light, noise and body. Perceptions are subjective and generate corresponding statements from employees, and these must be taken seriously. The complaints are often related to employees' ignorance and fear of losing something. The inner attitude to what is disturbing is decisive.

For example, if a person interested in airplanes lives in an approach lane, he has a positive attitude towards aviation and tolerates the volume of the passing airplanes; it is not disturbing for this person. A neighbour living nearby, who is rather negative towards aviation, will feel disturbed, because he lacks a connection to aviation.

Economy, health and well-being are not mutually exclusive. Over time, a greater awareness of the development and implementation of 'well-being' measures has developed. When employers let the idea and vision for more well-being grow and mature, employees see this as a sign of their employer's appreciation.

'Human resources' and related health and safety or sustainability management are the corporate functions that are responsible for the well-being of employees—creative measures and actions should emanate from these functions (well-being can also be triggered by sounding ancillary areas with music such as in certain zones—stairwell, toilet and foyer).

2.2 Air and Well-Being

2.2.1 Overview of the Chapter Air and Well-Being

The indoor climate is an essential prerequisite for well-being and comfort in rooms. In physical terms, a state of equilibrium between man and his environment must be established.

With food, humans absorb energy which they use in complex metabolic processes to maintain bodily functions, for example, muscle movement or the regulation of body temperature from 36 to 37 °C. During metabolism also metabolism energy is released, which the human being releases into the environment as heat. The energy conversions depend on the activity, but they also vary individually. Older people emit less energy than younger people.

Table 2.1 shows the human energy flows to be dissipated into the environment during various activities.

The metabolic functions require oxygen, which the human being inhales. Some of the oxygen is converted into carbon dioxide (CO_2), which is then released into the environment during exhalation. The physiological minimum of a person's air consumption is 4 l/s or 14.4 m^3/h.

When exhaling, not only CO_2, but also moisture is released into the air was delivered. The correct humidity of the inhaled air is also necessary for the functioning of the mucous membranes.

Together with the heating and cooling loads, the basic mechanisms of heat dissipation and respiration described above form the basis for the requirements placed on

Table 2.1 Energy output of humans during various activities [ISO 7730; cf. DIN EN 7730 (200G)]

Activity	Energy turnover W
Leaning	46
Sitting, relaxed	58
Sitting activity (office, apartment, school and laboratory)	70
Standing, light activity (shopping, laboratory and light industrial work)	93
Standing, medium–heavy activity (sales activity, housework and machine operation)	116
Walk on the level	
2 km/h	110
3 km/h	140
4 km/h	165
5 km/h	200

the human thermal environment. Health and well-being are essentially determined by the state of equilibrium of thermal comfort.

The thermal comfort can be determined by means of the environmental parameters

- room temperature,
- room air humidity,
- room airspeed.

Taking into account that the sizes,

- thermal resistance of clothing and
- activity level

Describe

The limits within which these parameters should lie are known and a variety of proven technologies are available to ensure compliance with the comfort parameters.

However, experience shows that compliance with the parameters does not necessarily guarantee the satisfaction and well-being of all room users. The dissatisfaction rate predicted percentage of dissatisfied (PPD) [cf. DIN EN 7730 (200G)] was therefore introduced into the overall assessment of a thermal indoor climate. Extensive research has shown that a PPD value of 6% cannot be undercut. Experience has shown that this value is not reached if the technical equipment is carefully planned and executed. The ISO standard 7730 [see DIN EN 7730 (200G)] works with the three PPD values 6, 10 and 15%. Thus, the quality of a room climate can be classified into the categories high–medium–low and is therefore also economically assessable.

In addition to thermal comfort, the odour quality, the hedonics, must also be included in the overall assessment of a room.

The individual comfort parameters are explained below. Finally, they are summarised in a checklist and assigned to the categories high–middle–low (Table 2.2). The checklist can be used as a decision-making basis for the necessary coordination processes between the room users and the client.

2.2.2   Room Temperature

During the heat period as well as the cooling period, the room temperature should remain within certain limits regardless of the outdoor temperature. The room temperature here is to be determined as the operative, i.e. felt room temperature. It is made up of the air temperature and the temperature of the room enclosing surfaces. In order to achieve low dissatisfaction rates, the temperature of the surrounding surfaces should be as uniform as possible.

Measurements have also shown that the room surface through which the temperature differences are generated is of considerable importance, as Tables 2.3 and 2.4 show.

The maximum permissible temperature differences to limit the dissatisfaction rate are specified for categories A, B and C [cf. DIN EN 7730 (200G)].

Table 2.2 Evaluation scheme for the evaluation of the indoor climate according to FGK-SR 17 (cf. FGK Status Report 8 2007)

No.	Criterion	Category			
	Category No	1	2	3	4
	Requirement level	High	Medium	Low	No A.
1.	Room temperatures and humidity				
1.1	Minimum room temperature heating period winter (clothing insulation 1 clo)	22 °C	21 °C	20 °C	<20 °C
1.2	Maximum room temperatures in winter	23 °C	24 °C	25 °C	>25 °C
1.3	Maximum room temperature cooling period/summer (reference value 33 °C fresh air temperature and 0.5 clo)	25 °C	26 °C	27 °C	>27 °C
1.4	Individual room temperature control refers to the control equipment and not to the design	±3 K	±2 K	Not possible	Not possible
1.5	Indoor humidity in winter	>40%	>30%	>20%	Unclassified
1.6	Indoor humidity in summer	<50%	<60%	<65%	Unclassified
1.7	Indoor humidity in summer	<11 g/kg	<12 g/kg	×Outside	Unclassified
2.	Fresh air volume flow				
2.1	Fresh air volume flow building m³/hm² qb	≥ 3,6	≥2,52	≥ 1,44	< 1,44
2.2	Outdoor air flow depending on person m³/h per person qP	≥ 36	≥25,2	≥ 14,4	< 14,4
2.3	Sum of 3.1 and 3.2 shall be used (occupancy of persons shall be specified)	$qges = n \times qp + A \times qb$			
3.	Technical plant boundary conditions				
3.1	Air filtration	IDA 1	IDA 2	IDA 3/4	Unclassified
3.2	Tightness of the air distribution system	B	C	D	Unclassified
4.	Further comfort parameters				
4.1	Draught rate DR Compliance with values according to section	10%	20%	30%	Unclassified
4.2	Warm or cold floor	19 °C ≤ Fußboden-temp. ≤ 29 °C			
4.3	Temperature gradient	2 K/m	3 K/m	4 K/m	Unclassified
4.4	Sound	≤35 dB, reference to DIN 4109			
4.5	Hedonic valuation according to VDI 3882 Part 2	+4 to >2	+2 to 0	<0 to −2	Unclassified

Table 2.3 Local discomfort due to vertical temperature differences and warm/cold soil

Category	Percentage of dissatisfied people PPD[a] (%)	Vertical air temperature difference[b] (°C)	Surface temperature range of the floor (°C)
A	<6	<2	19–29
B	<10	<3	19–29
C	<15	<4	17–31

[a]Applies to the thermal state of the body as a whole
[b]Temperature difference at 1.1 and 0.1 m above floor level

Table 2.4 Local discomfort due to radiation asymmetry

Category	Percentage of dissatisfied people PPD[a] (%)	Asymmetry of radiation temperature (°C)			
		Warm blanket (°C)	Cool wall (°C)	Cool ceiling (°C)	Warm wall (°C)
A	<6	<5	<10	<14	<23
B	<10	<5	<10	<14	<23
C	<15	<7	<13	<18	<35

[a]Applies to the thermal state of the body as a whole

Tables 2.3 and 2.4 show that cool ceilings and warm walls have the lowest dissatisfaction rates.

The subjective perception of temperature depends on individual preferences, clothing and activity. In usage units in which several people are staying, the room temperature requirements can, therefore, vary considerably in some cases. It should also be noted that the system technology can only guarantee the room temperature for heating and cooling within certain limits.

The statutory minimum requirement of the Workplace Ordinance for heating at 20 °C is probably too low for most room users. The value of 22 °C leads to considerably lower weight rates.

For the room temperature in summer, there is no legal maximum. In the Workplace Ordinance, 26 °C is mentioned as the target limit. Limitation of the room temperature in summer is only possible with cooling equipment or air conditioning. The regulations for air-conditioned offices assume a target value range of 25 °C (high requirements) to 27 °C (low requirements). If the reference outdoor temperature of 33 °C is exceeded, the room temperature rises above the specified values.

The individual intervention of the user is indispensable for the acceptance of the room temperature, and this applies especially in single offices or offices with few persons.

The plant engineering has to meet the requirement to create the smallest possible zoning of the heating and cooling systems.

2.2.3 Room Air Humidity

It is a well-founded finding of comfort research that human beings are able to maintain a temperature range between 21 and 22 °C and a relative humidity of from 40 to 50%. At low humidities, as they occur in winter, the mucous membranes dry out. They can only perform their task of retaining dirt and germs from the inhaled air to a limited extent. The development of respiratory diseases such as coughs, colds, bronchitis, etc. is promoted.

If the inhaled air is too dry, the body activates the entire airway nasal bronchial alveoli as a humidifying medium. The moisture that is not provided by the ambient air must be applied from the body itself and released into the respiratory air via the airways.

Figure 2.1 shows the human-biological interactions with room humidity (Scofield–Sterling diagram).

The relationships between the desired room air temperature and the room air humidity can be represented in a diagram, Fig. 2.2. The inner zone forms the area which is called the comfort area. The surrounding area is still considered comfortable.

Outside this range, the air conditions are uncomfortable; they are too humid or too dry.

At low outside temperatures, the air can absorb little moisture or water vapour. If the outside air is heated to room temperature, the relative humidity drops because the water content of the air is not changed. If, for example, air is heated from 0 °C/80%

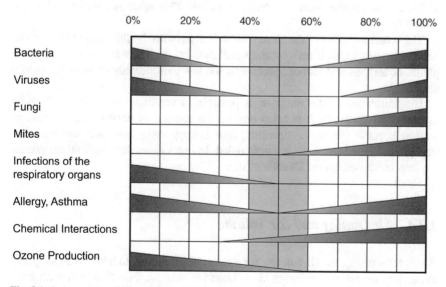

Fig. 2.1 Indoor air humidity and human-biological interactions according to Scofield and Sterling (1992), p. 52

Fig. 2.2 Comfort field. Relationships between room temperatures and room humidity

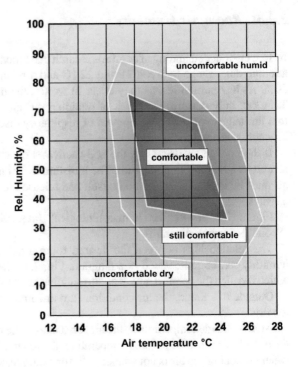

to 22 °C, the relative humidity drops to 18.5%. This value is uncomfortable and unhealthy.

The actual humidity of the air in a room is determined by the amount of moisture emitted by people, their activity and ventilation habits. The water vapour output of plants, aquariums and indoor fountains is too low to noticeably increase the indoor air humidity.

Humidification of the room air is possible in ventilation and air-conditioning systems; unfortunately, it is often omitted for reasons of energy and hygiene, as well as because the annual operating time is supposedly too short and therefore negligible. These arguments are unfounded. In many cases, humidification systems are then retrofitted at considerable cost.

2.2.4 Air Quality and Air Volume

When determining the air volume for a room or a building, in addition to the physiologically necessary minimum air volume for persons, two further criteria must be taken into account. These are the pollutants released in the room from building materials and the prepollution of fresh air with pollutants. In air-conditioning and ventilation technology, the term outdoor air is therefore used instead of fresh air.

Table 2.5 Evaluation of the CO_2 content of indoor air by the Federal Environment Agency (cf. "Health assessment of carbon dioxide in indoor air" 2008)

CO_2 concentration	Evaluation of the UBA
<1000 ppm	Harmless
>1000–2000 ppm	Conspicuous
>2000 ppm	Unacceptably

Table 2.6 Evaluation of the CO_2 concentration in the room according to EN 15251 (2007)

Cat. acc. to EN 15251	CO_2 concentration in the room	Expected dissatisfaction rate (%)
I	Outdoor air concentration + 350 ppm	15
II	Outside air concentration + 500 ppm	20
III	Outside air concentration + 800 ppm	30

The CO_2 concentration of the outside air is currently 400 ppm

The first criterion is the CO_2 content of the air. CO_2 is an odourless and invisible gas which, in higher concentrations, can cause headaches, fatigue, and even aches and pains. dizziness and lack of concentration leads. Table 2.5 shows the Federal Environment Agency's assessment of the CO_2 concentration in a room.

The CO_2 concentration of the outside air is currently 400 ppm. Studies of the dissatisfaction rate at different room air concentrations are shown in Table 2.6.

A comparison of Table 2.5 with Table 2.6 shows that only Categories I and II can be used if dissatisfaction rates <20% are targeted.

In the course of harmonising European standardisation, considerably higher CO_2 concentrations are sometimes categorised.

The second criterion for the outdoor air rates of rooms is the removal of the pollutants that are produced in the rooms.

Volatile organic compounds (VOC) are vapours from building materials. These include hydrocarbons, alcohols, aldehydes and organic acids. Solvents and organic compounds from biological processes are also among the VOCs. Unfortunately, there is no strictly verifiable relationship between concentration and effect for VOC mixtures. For this reason, the categorisation by the UBA according to Table 2.7 is to be understood as a hygienic precautionary area (cf. Seifert 1999).

In the relevant standards (cf. EN 1525 2007), the amount of fresh air required to dissipate the pollutants is taken into account in relation to the area. The air quantity for creating comfortable indoor air conditions is, therefore, the sum of the breathing air requirement of the persons and the area-dependent air quantity for the removal of pollutants.

Table 2.8 contains an example calculation for an office building with the occupancy of 10 m^2/person.

Table 2.7 Categorisation of VOC concentrations by UBA (Cf. Seifert 1999)

VOC concentration	UBA rating
0.2–0.3 mg/m^3	Target value for the long-term average
1–3 mg/m^3	Target value for rooms for a long-term stay
10–25 mg/m^3	Value can be reasonably expected on a daily basis only temporarily, if at all
From 8 mg/m^3	Irritations to eye and nose possible
From 25 mg/m^3	Inflammatory reactions and limitations of lung function possible

Table 2.8 Calculation example for the required air volumes for non-residential buildings with low-pollutant emissions

Category according to EN 15251	Person-related air volume (m^3/h)	Air volume for NWGs 10 m^2/person and low-pollution building (m^3/h)	Sum of the air volumes = required air volume for 10 m^2 (m^3/h)	Area-related air volume (m^3/hm^2)	Expected percentage of unsatisfied employees (%)
I	36.0	36.0	72.0	7.2	15
II	25.0	25.2	50.4	5.04	20
III	14.0	14.4	28.8	2.88	30

2.2.5 Window Ventilation or Ventilation System

The openable window has a very high significance for office users as a means of influencing the local workplace climate, even if opening the window is occasionally associated with too high or too low temperatures, draughts and a reduced increase in external noise. Apparently, the office user is ready to make compromises regarding comfort for the individual window opening option.

The air supply through openable windows is only effective in the outer zone of buildings up to a distance of approx. 5.5 m from the facade. In deep rooms, there is a risk that the air supply in the inner zone will be insufficient despite the windows being open, but that complaints about inadequate comfort will be expressed in the outer zone.

The Workplace Ordinance (cf. ArbStättV 2004) and the Workplace Directive ASR 5 (cf. ASR A3.6 2012) specify criteria and limit values for the room depth to which so-called free ventilation via windows is possible and the size of the opening area of the windows for continuous ventilation and so-called shock ventilation, i.e. the simultaneous opening of all available windows, see Table 2.9.

In modern office buildings with a large building depth, dense occupancy and variable zone or room layout, air supply and disposal can only be solved satisfactorily with a mechanical ventilation system. This refers not only to the amount of air exchanged, but also to the uniform distribution of air. Finally, it should be noted

Table 2.9 Limits of window ventilation according to the Workplace Ordinance

Ventilation system	Criterion 1	Criterion 2	Criterion 3
	Maximum permissible room depth in relation to clear room height (LRH) in m	Opening area to ensure minimum air exchange	
		For continuous ventilation in m² per person present	For shock ventilation in m² per 10 m² floor area
One-sided ventilation	Maximum room depth = 2.5 × LRH (for LRH > 4 m: max. room depth 10 m)	0.35	1.05
Transverse ventilation	Maximum room depth = 5.0 × LRH (for LRH > 4 m: max. room depth 20 m)	0.2	0.6

The indicated opening areas are the sum of supply and return air areas

that air exchange with a ventilation system is much more energy efficient due to the possibility of heat recovery.

2.2.6 Air Velocity

The sensation of draughts in offices is only addressed in connection with ventilation systems, although rooms that are only ventilated and vented via openable windows can sometimes experience considerable draughts.

Draughts are due to excessive air velocity. Further decisive influencing variables on the draft sensation are the room air temperature and the speed fluctuations, for which the turbulence degree is the measured variable. Figure 2.3 shows the relationships between mean air velocity, air temperature and turbulence degree for the draft risk 10% (DR 10% = draft risk).

It can be seen that average air velocities below 0.20 m/s are unproblematic.

2.2.7 Heating and Cooling Loads

The creation of a thermal room environment that is beneficial to health must be carried out by the heating system—and cooling load the building and the rooms.

A distinction must be made here between external and internal loads. The external loads arise from the exchange of energy through the building envelope. The internal loads are the sum of the quantities of waste heat from computers, monitors, printers, lighting and also the heat emitted by humans.

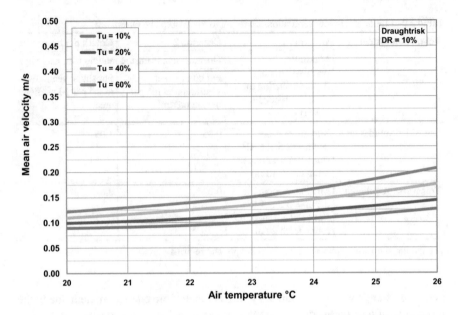

Fig. 2.3 Maximum air velocity as a function of the local air temperature and the degree of turbulence (Tu) according to DIN ISO 7730—Category A corresponds to DR = 10%

The heating load consists of the following components: Transmission heat demand and ventilation heat demand. Transmission heat demand refers to the heat loss through the building envelope in winter. Due to the legal requirements for thermal insulation, the heat losses via the building envelope have been considerably reduced, and this applies in particular to the windows.

A significant side effect of the reduced heat loss is the higher surface temperature on the inside of walls and windows. For example, the interior temperature of modern window glazing no longer drops below 18 °C at very low outside temperatures. In the insulation standard of the 1980s, this value was still 12 °C.

The ventilation heat requirement arises from the need to heat the outside air required to supply the rooms with air to room temperature. In rooms without a ventilation system, the radiators must be dimensioned accordingly; in ventilation systems, the outside air is heated in the ventilation unit. By using heat recovery systems, the ventilation heat requirement can be provided more efficiently than with window ventilation.

Solar radiation is the main factor influencing the exchange of energy between the room and its surroundings. In winter, the incident radiant energy through the windows can support and temporarily replace the function of the radiators. In extreme cases, however, excessive heating takes place so that there is a cooling case. This can be remedied by using effective external sun protection.

The incidence of solar radiation represents the largest part of the external cooling load in summer. The heat incidence through the outer walls is negligible. An exception is a roof. The energy input through the roof must be taken into account when considering the cooling load of the top floor.

The power requirement and thus the waste heat from modern office equipment have fallen sharply in recent years. This is not only a contribution to energy efficiency, but also makes it easier to create a comfortable room. In the last ten years, technical progress has halved the amount of waste heat from 15 W/m^2 to approx. 7.5 W/m^2, with a further downward trend. The situation is similar to lighting. As a result of the use of LED technology, the waste heat has been reduced from 12 to 8 W/m^2 with a downward trend. As can be seen from Table 2.1, a person emits 70 W when sitting in an office; the occupancy density of 10 m^2/person results in 7 W/m^2.

The total internal cooling load of 22.5 W/m^2 is calculated from the above numerical values. If one adds to this value the external cooling load, which for modern facades is approx. 25 W/m^2, related to the floor area of a workplace of 10 m^2, a total cooling load of approx. 47.5 W/m^2 is obtained.

With densify occupancy concept, if the cooling load is 8 m^2/person, the internal cooling load would increase from 22.5 to 28 W/m^2 and the total cooling load to 53 W/m^2.

With these loads and the compression of workplaces, window ventilation in larger rooms is no longer sufficient to produce uniform air quality at all points in the room. A ventilation system with the cooling function must be used here.

Air–water systems are mainly used here for office buildings. With these systems, comfortable room air conditions can be created with cooling loads of up to 80 W/m^2. The use of surface cooling systems, e.g. cooling ceilings, is particularly advantageous, n, where most of the heat transfer takes place by radiation, so that the room air velocities are low. The air is needed for air exchange and humidity regulation.

Room temperature limitation in summer is possible with ventilation and air-conditioning systems with cooling function. However, care should be taken to ensure that the systems not only cool, but also dehumidify. In the summer of 2016, the so-called sultriness limit of the outside air, at which the water content of the air is greater than 9.6 g water per m^3 air, was exceeded for 455 h.

2.2.8 Hedonic Assessment of Indoor Air Quality

In the previous sections, the comfort in rooms was related to physical and technical parameters that are maintained with suitable systems for heating, cooling, humidification, dehumidification and air purification. Nevertheless, there are occasional complaints about 'artificial air' or general complaints about discomfort, which cannot be explained by a deviation from the parameters of thermal comfort.

This is where Hedonik (=odour quality), which has to be considered parallel to the thermal comfort. The scale for hedonics is a 9-digit rating scale that ranges from −4 (=unpleasant) to 0 (=neutral) to +4 (=pleasant). Hedonics is perceived individually and in some cases very differently; its social origin and environmental influence are important influencing factors that can also change over time. In addition, measured

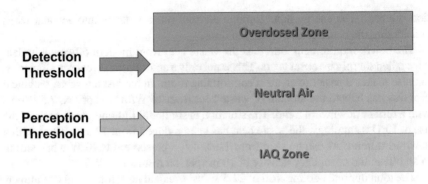

Fig. 2.4 Dosage of fragrances according to the concept of fragrance technology

identical air concentrations are perceived differently by test persons. Temperature and humidity also influence the hedonic effect.

Humans are directly exposed to olfactory sensations because the olfactory receptors are located in the mucous membranes of the respiratory tract and any olfactory stimulus is directly registered by the brain.

The hedonics can be improved by adding additives to the air (cf. Kempski and Ziegler 2000).

The addition of more or less perceptible fragrances to the air by stand-alone devices in the room is not necessarily effective.

Another approach is pursued by fragrance technology. Substances are developed from the combination of fragrances, with which one tries to synthesise the substances contained in the natural air. The substances are then preferably distributed through the ventilation system in the room. The dosing moves between the perception and detection thresholds, see Fig. 2.4.

The hedonic evaluation of the indoor air can be carried out according to VDI Guideline 3882, Parts 1 and 2. The 9-digit scale described above is used for the evaluation.

A questionnaire is presented to a number of test persons in which details of the odour assessment are asked. The evaluation of the questions is entered in the 9-digit scale. The answers are then evaluated using statistical methods. Positive ratings on the hedonic scale range between +2 and +4.

It should also be noted that methods of odour assessment are state of the art in the industry. Examples include the food industry, the automotive industry and components and products for the construction industry. The range of applications extends from the detection of harmful vapours (cf. DIN ISO 1600-28 2012) to the conscious scenting of products to produce a product-specific, acceptable odour sensation. The application in room air technology is only one of the numerous fields of application.

2.2.9   Evaluation of the Indoor Climate

The level of requirements for the air supply and the technical equipment depends on the wishes of the users and the client. Economic reasons are also of great importance. It is strongly recommended to create the greatest possible transparency about the project objective before the start of planning, in consultation with all parties involved.

The following is an evaluation scheme for the indoor climate (cf. FGK Status Report 8 2007). The desired category is marked with the criteria high, medium and low (=category 1–3) and category 4 (=not classified).

The evaluation scheme applies to buildings with ventilation or air conditioning systems. In buildings with window ventilation, only parameters 1.1, 1.2, 5.4 and 5.5 of room temperature and ventilation can be guaranteed.

2.3   Din

2.3.1   Healthy Acoustics in Office and Administration Buildings

New construction, conversion, refurbishment, technical adaptation and operation of office and administration buildings are complex undertakings that require sophisticated and comprehensive planning (Fig. 2.5). The building physics aspects of acoustics such as facade soundproofing or soundproofing to foreign areas, noise levels of building services installations, etc. are regulated by the specifications of DIN 4109—anchored in every state building law—under building law. Compliance checks are usually carried out by the authorities in connection with compliance with the requirements of the building application. Sound insulation, also known as building acoustics, is assigned to the term 'technology' in Fig. 2.5.

In Fig. 2.5, room acoustics is assigned to the ergonomics area. Ergonomics (Fig. 2.6) summarises the most important parameters of the health-relevant influencing variables in an office building. Each subarea requires in itself a maximum of balance and harmony. Even slight deviations in the individual disciplines have an enormous effect on the well-being, health and motivation of the employees.

An unsuitable, not appropriate work chair (s. Chap. 4) (anthropometry) leads to back pain and thus automatically to the rejection of the remaining ergonomic factors; the performance drops drastically.

The design elements from Fig. 2.7 room acoustics/sound environment are available to us for the subarea of room acoustics/sound environment from ergonomics (Fig. 2.6).

Since room acoustics If the acoustic and acoustic environment have a very strong influence on the health, well-being and motivation of office workers, this area is examined in detail here and described with the so-called "**ABC of** office acoustics".

A for absorb, absorb sound (Fig. 2.8).

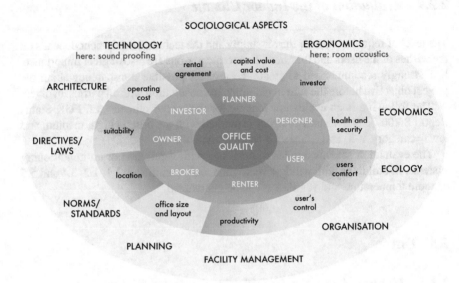

Fig. 2.5 Office quality

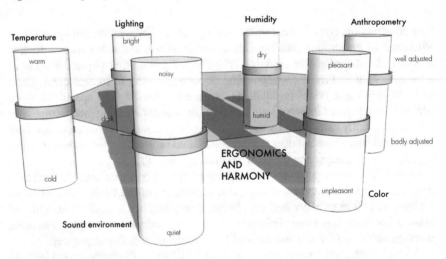

Fig. 2.6 Ergonomics

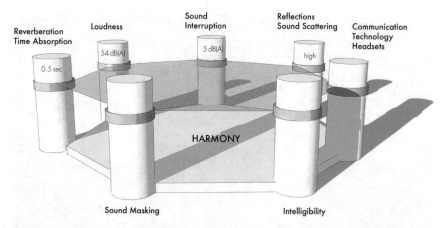

Fig. 2.7 Room acoustics/sound environment

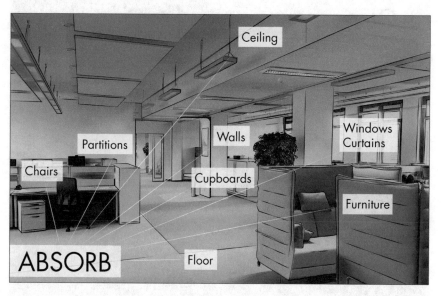

Fig. 2.8 Absorbent

B for block, interrupt sound and scatter (Fig. 2.9).

C for cover, mask (Fig. 2.10).

In Fig. 2.7, the absorb area comprises the columns absorption/reverberation time and partly also the volume.

In Fig. 2.7, the block section covers the columns interrupted sound and reflection/scattering sound as well as partly the volume.

Fig. 2.9 Block

Fig. 2.10 Cover

The cover section in Fig. 2.7 covers the pillars of comprehensibility of words and sound masking and deals with the intentional or unintentional comprehensibility of conversations between colleagues.

In the second step, the individual columns from Fig. 2.7 room acoustics/sound environment and their effect are examined.

2.3.2 Column: Reverberation Time/Absorption

The reverberation time and thus the acoustic comfort in a room are largely determined by the sound absorption present (see Fig. 2.11). This influences the volume of the background noise and the degree of speech intelligibility.

The sound absorption coefficient of a material describes its ability to extract energy from an incident sound field. It is given in the range from 0 (no absorption, total reflection) to 1 (total absorption, no reflection), e.g. a material with an absorption coefficient of 0.5 withdraws 50% of its sound energy from the sound field.

Figure 2.12 shows the effect of sound-absorbing surfaces. Figure 2.8 shows possible absorption areas such as ceiling, floor, walls, windows/curtains, chairs, partitions, cupboards and furniture.

As a rule of thumb it can be assumed that an office room is acoustically well conditioned if approx. one third of the entire room and furniture surfaces are 100% absorbent (sound-absorbing). It is important that the absorbing/sound-absorbing surfaces are evenly distributed over the existing surfaces.

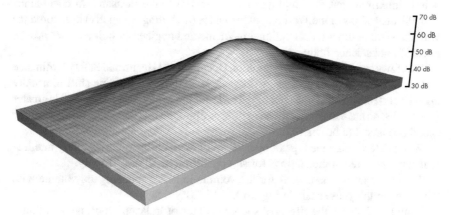

Fig. 2.11 Hallowing room without absorption

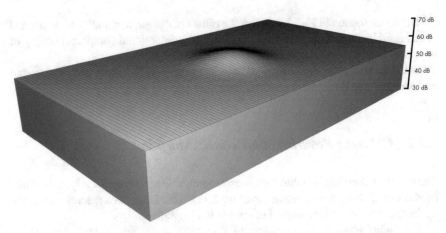

Fig. 2.12 Pleasant quiet room with absorbent ceiling and carpeting

2.3.3 Column: Volume and Speech Style

To the volume says the Workplace Ordinance in the appendix under point 3.7:

'**In the workplace, n is the sound pressure level as low as possible, depending on the type of operation**. The sound pressure level at the workplace and in work rooms shall be reduced, depending on use and the activities to be performed, to such an extent that there is no adverse effect on the health of employees'.

Concrete limit values are not mentioned in the Workplace Ordinance. However, it is pointed out that the employer must take protective measures in accordance with the provisions of the Workplace Ordinance (including Appendix) in accordance with the state of the art, occupational medicine and hygiene as well as other proven occupational science findings.

It follows from the Noise and Vibration Occupational Health and Safety Ordinance that an assessment level (average workplace level during a working shift of usually eight hours) of 80 dB(A) must not be exceeded at the workplace. Rating levels greater than 80 dB(A) are usually not given in offices, although this fact alone does not allow 'good acoustics' to be derived.

When determining the type of operation, for example as 'predominantly intellectual activities', recommendations for an upper limit of 55 dB(A) apply.

A very important parameter for the volume in office rooms is the volume with which the employees speak during work.

Figure 2.13 shows the effects of speech patterns of 'relaxed', 'normal' and 'loud' in an acoustically well-conditioned room. A relaxed way of speaking is always recommended for voice health, volume in the room as well as for a pleasant effect of speech on the conversation partner.

Examples of different speech volumes are shown in Fig. 2.14 for the recommended relaxed speech style and for the 6 dB(A) louder speech style called 'normal'. On the volume scales to the right of the respective map representations of the computer

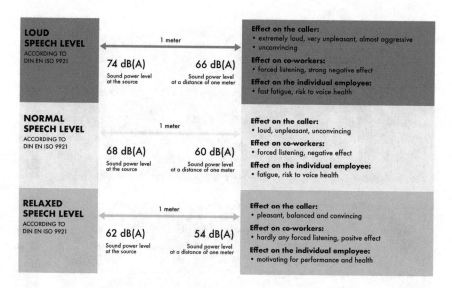

Fig. 2.13 Way of speaking

simulations, the clear differences in volume can be read. In order to reduce the volume in the louder room from 56 dB(A) to 50 dB(A) in terms of building physics, the existing absorption area in the room would have to be quadrupled, which is not possible because the necessary surfaces are simply not available.

2.3.4   Column: Sound Interruption

The subject of sound interruption /shielding is illustrated schematically in Fig. 2.15. The height to be used and the absorption capacity of shields to be used (e.g. highly absorbent or transparent/translucent) depend on the activities carried out in the room. Room surfaces also have a considerable influence on shielding performance; a highly absorbent ceiling produces a better shielding performance than a low absorbent ceiling. Furthermore, an accounting department and a lively sales office differ considerably in volume and sound. And also the specific room acoustic sensitivities of the employees in the room are directly related to the concentration requirements of the respective activities.

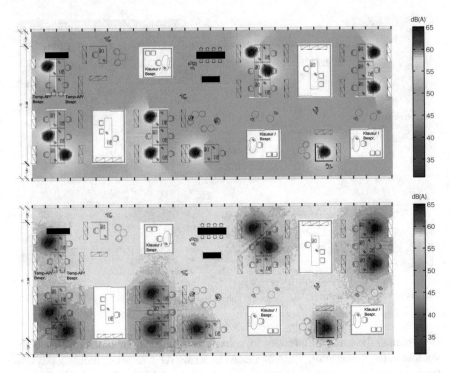

Fig. 2.14 Volumes in the office

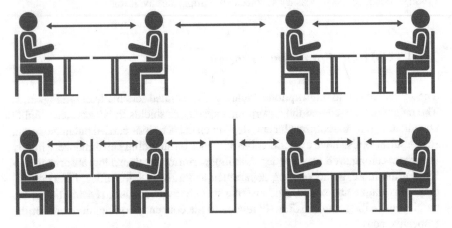

Fig. 2.15 Sound interruption

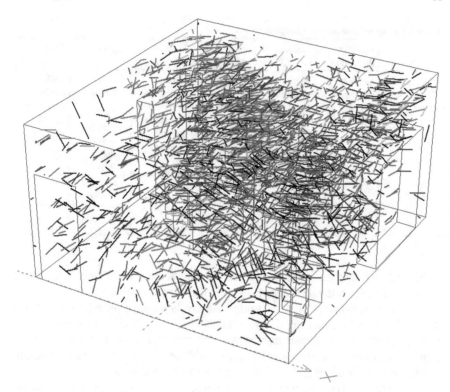

Fig. 2.16 Reflection/sound scattering

2.3.5   Column: Reflection/Sound Scattering

Figure 2.16 shows the reflection of the/sound scattering as the subset of the sound energy reflected by the room boundary surfaces (walls/ceiling/windows etc.) and all furniture and objects in the room and scattered according to their geometric shape. The reflected and scattered sound energy together with the direct sound (e.g. speaker) produces the volume in the room. The way the employees speak determines the volume in the room significantly.

2.3.6   Pillar: Communication Technology/Headsets

Telecommunications have become increasingly important in companies. Personal contact with customers has declined in many business areas. Today, customer contacts take place via the Internet, telecommunications or e-mail. In telecommunications, a balanced, pleasant sounding, sympathetic and convincing voice is an important basis for success.

Table 2.10 Requirements for modern headsets

Sound quality	Hearing protection	Comfort	Safety and security	Speech volume	Ergonomics	Data protection
Individually adjustable sound	Hearing protection at noise peaks	The volume of incoming calls is automatically adjusted	Background noise for incoming calls is reduced	Muted speech is sufficient. The voice remains healthy and a balanced voice convinces the customers	– Free hands – Mobility – No neck tension due to telephone receiver	Callers cannot hear room or call content

Table 2.10 shows the recommendations for modern headsets. The majority of the devices available on the market already meet these requirements today.

The most important point when using headsets, however, is the adjustment of the devices to the individual user. Everyone hears differently; hearing ability decreases with age and ambient sounds also have a strong influence on hearing. Individual setting options in the headsets can be very helpful here. When procuring such equipment, attention should therefore also be paid to such possibilities.

A frequent misuse, for example, is the incorrect positioning of the speech microphone (speech tube); the microphone is too far away from the mouth and/or points into the room. The result is that the highly sensitive directional microphone records the user's language and room noises (conversations with colleagues) and transmits them to the caller. The result of the wrong microphone position inevitably results in ever louder speaking in order to drown out the surrounding colleagues' conversations.

A further factor is the increase in volume in the headset, which in turn leads to even louder speech. Once the volume has been turned up, it is often left at that level, which means that when you start working well rested and rested the next day, you immediately start speaking loudly because you hear loudly.

If the effectiveness of the directional microphone in the headset is impaired by incorrect microphone positioning and room calls are transmitted to the caller, this is not only an acoustic problem, but also a data protection problem.

Suitable equipment and correct use therefore guarantee acoustic satisfaction.

2.3.7 Pillar: Word Comprehensibility

As shown in Fig. 2.17, the ambient and working noises in offices have almost disappeared in recent years. Office buildings are now well acoustically protected against noise from outside, from ancillary rooms, from floors above or below due to legal reg-

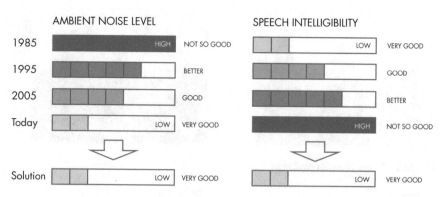

Fig. 2.17 Word comprehensibility

ulations, so the general ambient noise is very low. Office equipment has also become quieter due to corresponding legal requirements. What have been kept are the speaking, acting persons, who—precisely because of the lack of ambient noise—are now well understood at every point in the room.

Calm has become a problem

How solutions can be found here is explained in the following using the column sound masking.

2.3.8 Influence and Effect of Sound Masking on Speech Intelligibility

It is common knowledge that in a quiet environment speaking people are understood more clearly. As soon as it gets louder in the surroundings, for example, in the restaurant or in the airplane, we hardly understand even our direct neighbours any more. Ambient noises of this kind are, however, not accepted in the volume and sound of office workers, although it is almost impossible to hear and be heard in noisy environments. Also, noises from air-conditioning systems or tilted windows for the admission of street noises are not very well received, because then one knows where the noises come from, and thus acoustic assignability of the noise and consequently an actual, additional source of interference is given.

It is therefore important to find satisfactory opportunities for office workers. One possible solution is shown by the following three graphs with cause representation (Fig. 2.18) and representation of the effect and influence of sound masking on speech intelligibility.

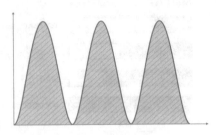

Fig. 2.18 Causes of sound masking

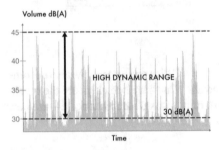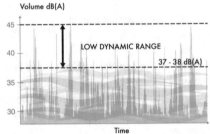

Fig. 2.19 Measure for sound masking

Cause

It is, therefore, necessary to find a system on the market which brings a pleasant, unassignable and least possible disturbing carpet of sound into the room, but which to the greatest possible extent prevents distraction through understandable background language (and thus also one's own being heard) (Fig. 2.19).

Measure

Employees should no longer find such a system (which is preferably based on sounds of nature familiar to humans) disturbing after a period of familiarisation during routine work. A prerequisite for the successful application of such systems is the approval of innovative techniques by all parties involved. A successful basis for this can be provided by well-managed change management.

Effect

The effectiveness of sound masking systems can be very well demonstrated with a computer simulation, taking into account all real acoustic room conditions (Fig. 2.20).

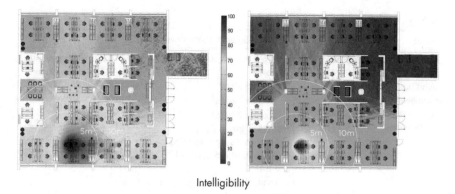

Intelligibility

Fig. 2.20 Effect of sound masking

2.3.9 Standards—An Overview

According to the room acoustic design possibilities, the following tables should also take into account the diversity of existing standards. for reverberation time (Table 2.11), level decrease per distance doubling (Table 2.12), sound pressure level (Table 2.13), speech level (Table 2.14) and background noise level (Table 2.15).

2.3.10 Standards—A Critique

The contents of the ASR ArbeitsStättenRichtlinien A3.7—Draft 01/2016—become legally binding regulations after adoption and publication as part of the Arbeitsstättenverordnung.

All other standards listed contain room and building acoustics recommendations, which can become requirements through private law agreements. Some standards, in turn, correspond to the so-called recognised rules of technology, to which, in turn, the Workplace Ordinance and the Occupational Health and Safety Act refer.

For new construction, conversion, technical adaptation and operation of office and administration buildings, a requirement profile—professionally accompanied—should therefore always be drawn up. If there are works councils or staff councils in companies, it is always advisable to check compliance with information and co-determination obligations in accordance with the Works Constitution Act. Early involvement of works councils or staff councils usually leads to greater consensus and better results. In this context, it should also be examined whether company agreements between companies and staff representatives have already been agreed in advance, whether their contents conform to the planned construction projects or whether adjustments are necessary.

Table 2.11 Reverberation time

Norm	Application area, room type	Value
DIN 18041 (new edition March 2016) 'Audibility in rooms—requirements, recommendations and notes for planning'	Group A rooms, e.g. meeting rooms: – Volume 60 m^3 – Volume 120 m^3 – Volume 250 m^3 Rooms in group B, e.g. Call centre (usage type B5) Open-plan offices (type of use B4) Individual offices (usage type B3)	0.40 s (0.32 with inclusion) 0.50 s (0.40 with inclusion) 0.60 s (0.48 with inclusion) 0.54 s (derived from $A/V \geq 0.30$) 0.65 s (derived from $A/V \geq 0.25$) 0.82 s (derived from $A/V \geq 0.20$)
VDI 2569 current (Edition January 1990) 'Sound insulation and acoustic design in the office'	Small offices Large offices	No requirement 0.45–0.55 s
VDI 2569 new (published draft February 2016)	Room acoustics class A: call centre recommendation Room acoustics class B: recommendation design office or similar, minimum recommendation call centre Room acoustics class C: minimum recommendation for design office or similar	0,60 s 0.70 s (single office: 0.80 s) 0.90 s (single office: 1.00 s)
DIN EN ISO 11690-1 (1996) 'Guidelines for the design of low-noise workplaces equipped with machinery—part 1: general principles'	General workspaces – Volume <200 m^3 – Volume 200–1000 m^3 – Volume >1000 m^3	0.5–0.8 s 0.8–1.3 s Only request for level decrease
DIN EN ISO 9241-6 (1999) 'Ergonomics requirements for office work with display screen equipment—part 6: principles for the working environment'	Offices – Volume 100 m^3 – Volume 500 m^3 – Volume 1000 m^3	Conducting conversations Otherwise. Usage 0.45–0.80 0.70–1.10 0.80–1.20
ASR A3.7 (draft 01/2016) 'Technical rules for workplaces—noise'	Call centre Open-plan offices Individual offices	0.5 s 0.6 s 0.8 s

Table 2.12 Level decrease per distance doubling

Norm	Application area, room type	Value
VDI 2569 new (published draft Feb. 2016)	Room acoustics class A: call centre recommendation Room acoustics class B: recommendation design office or similar, minimum recommendation call centre Room acoustics class C: minimum recommendation for design office or similar.	2/3 of the measuring paths: 8 dB Other: 6 dB 2/3 of the measuring paths: 6 dB Other: 4 dB 1/3 of the measuring paths: 6 dB Other: 4 dB
DIN EN ISO 3382-3 (05/2012) 'Measurement of room acoustic parameters—part 3: open-plan offices'	Open-plan offices	7 dB
DIN EN ISO 11690-1 (1996)	Working rooms in general, volume > 1000 m^3	3–4 dB
DIN EN ISO 9241-6 (07/1996)	Offices	4–5 dB
VDI 3760 (draft 01/2016) 'Calculation and measurement of sound propagation in working spaces'	Offices > 500 m^3 with subdivision by partition walls	4–5 dB
ASR A3.7 (draft 01/2016)	–	–

The existence and use of this diversity of standards in combination with 'ancient standards' and unfinished new drafts require considerable technical expertise in drawing up an individual requirements profile with the corresponding potential for future adaptation with the approval of existing employee bodies.

Acoustic Design for Offices—Conclusion

For room acoustic overall design, a remarkable statement from the draft of VDI 2569 02/2016, page 6, should be mentioned here; there—as follows—is explained:

30–40% of the nuisance effect from noise can be explained by technical-acoustic factors.

60–70% are due to so-called moderators of harassment.

Among the personal and situational moderators of harassment are the following factors:

- Control of the noise,
- Setting for the noise source,
- Predictability of the sound event,
- Activity profile of the employee,
- Organisational and corporate structure, with identification with the company,

Table 2.13 Sound pressure level

Norm	Application area, room type	Value
VDI 2569 current (edition January 1990)	Depending on the activity	Reference to VDI 2058-3
VDI 2058-3 (new edition Aug. 2014) 'Assessment of noise at the workplace taking into account different activities'	For predominantly intellectual activities	55 dB(A)
DIN EN ISO 9241-6 (1999)	Depending on 'difficulty and complexity'	35–55 dB(A)
VDI 2569 new (published draft February 2016)	Room acoustics class A: call centre recommendation Room acoustics class B: recommendation design office or similar, minimum recommendation call centre Room acoustics class C: minimum recommendation for design office or similar	2/3 of the measuring paths: 47 dB Other: 49 dB 2/3 of the measuring paths: 49 dB Other: 51 dB 1/3 of the measuring paths: 49 dB Remaining: 51 dB
DIN EN ISO 3382-3 (05/2012)	Open-plan offices	48 dB
DIN EN ISO 11690-1 (1996)	For routine office work for activities requiring concentration	45–55 dB(A) 35–45 dB(A)
baua Advice on risk assessment	Simple and predominantly routine tasks intellectual activities	45–55 dB(A) 35–45 dB(A)
ASR A3.7 (draft 01/2016)	Activities with high mental demands	55 dB(A)

- Workload, other environmental factors such as lighting and thermal comfort, and individual noise sensitivity."

The draft of VDI 2569 02/2016 is limited—as also noted on page 6—to noise abatement through building and room acoustic measures, i.e. to the mentioned approx. 30–40% of the nuisance effect.

However, the draft also points out that the human factors listed must at least be taken into account when planning office space. The major part of the nuisance effect of 60–70% is thus attributed to the human and design factors in the draft of VDI 2569 02/2016 (Fig. 2.21).

From this it can be concluded that successful room acoustics can only be achieved by considering the following points:

- **30–40%**
- Fulfilment of technical-acoustic factors
- **60–70%**

Table 2.14 Speaker level

Norm	Application area, room type	Value
DIN 18041 (new edition March 2016)	General	Relaxed speech: 54 dB(A) Normal speech: 60 dB(A)
VDI 2569 current (issue January 1990)	General	Relaxed speech: 54 dB(A) Normal speech: 60 dB(A)
DIN EN ISO 9241-6 (1999)	General	Relaxed speech: 54 dB(A) Normal speech: 60 dB(A)
DIN EN ISO 9921 "Ergonomics—assessment of speech communication"	General	Relaxed speech: 54 dB(A) Normal speech: 60 dB(A)
ANSI S3.5-1997 "Methods for calculation of the speech intelligibility index"	General	Normal speech: 60 dB(A)
DIN EN ISO 3382-3 (05/2012)	To be used for calculations according to this standard	Normal speech: 57.4 dB(A)
BGI 5141 (12/2012) "Acoustics in the office—aids for the acoustic design of offices" (Ed.: VBG—your statutory accident insurance)	Used for calculations in this study	Sound power level Lw = 63 dB(A), referred to as 'normal speech'
ASR A3.7 (draft 01/2016)	General	Colloquial language: 55–65 dB(A)

Fig. 2.21 Conclusion: overshot the target!

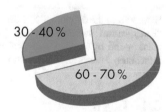

- Inclusion of design factors from Fig. 2.6, ergonomics, such as lighting and thermal comfort,
- Consideration of speech behaviour in connection with telecommunications technology,
- Development of rules of the game on the user's own responsibility (see Chap. 1),
- Implementation of change management geared to the requirements (see Sect. 2.1.1),
- Inclusion of the human resources department.

Table 2.15 Background noise level

Norm	Application area, room type	Value
DIN 4109 (07/2016)	All classrooms and workrooms	35 dB(A) for ventilation systems up to 40 dB(A), provided that it is continuous noise without conspicuous individual tones
DIN 18041 (new edition March 2016)	Category A rooms (e.g. conference rooms)	35 dB(A)
VDI 2569 current (issue January 1990)	Individual offices Multi-person offices	30–35 dB(A) 30–40 dB(A), if useful for masking up to 45 dB(A)
VDI 2569 new (published draft February 2016)	Room acoustics class A: call centre recommendation Room acoustics class B: recommendation design office or similar, minimum recommendation call centre Room acoustics class C: minimum recommendation for design office or similar	30 dB(A) 35 dB(A) 40 dB(A)
DIN EN ISO 11690-1 (1996)	Conference room Individual offices Open-plan offices	30–35 dB(A) 30–40 dB(A) 35–45 dB(A)
ASR A3.7 (draft 01/2016)	Conference room Individual offices Open-plan offices	35 dB(A) 40 dB(A) 45 dB(A)

Extract from the definitions of ASR A3.7, draft 01/2016, point 3.6

"**Background noises**" are noises in a room caused from outside—e.g. by traffic, production noises—and by installed technical equipment, without the noise sources occurring in the room (machines, devices and conversations with neighbours)

2.4 Light

2.4.1 Introduction to the Chapter

It should be out of the question: Daylight was there before man. The sun is our life-giver. Without light, we would not be here. Man has developed in the light and not the daylight with man. With artificial light, it is fundamentally different. Even in the second decade of the twenty-first century, artificial light, conceived by humans and made for humans, is still in the process of adapting to humans. This means first and foremost his physiological and psychological needs in order to enable a person to act in a given environment.

With the use of LED light sources, the development of artificial light has experienced an acceleration of technical and creative use, which nobody thought possible just five years ago. The lighting industry and many lighting engineers would certainly have preferred a slower development, but at the moment they are driven by user wishes that welcome and demand the positive effects of new artificial lighting technology.

The perception of light to the visual system of the eyes is a very simplified view. Not only the eyes, but also the skin and the hair process light and radiation. Good light does not only mean good visual performance, but it can also, if used correctly, positively influence our entire organism. Daylight awakens, the sleep hormone melatonin is inhibited. Sunlight, i.e. radiation light, releases the 'happiness hormone' serotonin in the brain and promotes vitamin production, in particular, vitamin D3.

Darkness means less light, but also relaxation and calm. Darkness can be experienced positively, darkness not. In darkness, we cannot see the things we want to see. This is usually caused by glare from other light sources in the visual field. In the ecological environment, this can lead to light pollution by unblinded light source n be, in the office a window without glare protection.

The following pages of this section 'Lighting in the office' are intended to help illustrate the basic relationships between lighting in the office and to justify basic decisions in lighting design for office activities.

2.4.2 Light and People

Basically, you see between physiological and psychological processes. Another distinguishing criterion is the visual and non-visual and biological effects of light on humans.

Until a few years ago, the latter was only treated by biology and medicine. This changed at the beginning of this millennium, when a "third photoreceptor" was identified without a doubt in the eye of our retina, i.e. not only suppositories (receptor The receptors (receptors for light/dark vision) and the receptors (receptors for colour vision) are present. This newly discovered receptor controls biological light effects on the nervous system and is also responsible for the hormonal secretion and inhibition processes. Also 'no light' is of psychological and physiological importance!

The surrounding lighting must ensure the functioning of the visual system, i.e. the physiological process in the eye and brain. This is necessary, if not sufficient, condition of light and action, whether in the office or elsewhere.

The eye is able to provide visual performance in every daylight and nightlight situation. The human eye has the ability to adapt to various illuminances. The extent of adaptation ranges from 0.1 lx in a starry night to daylight, which reaches from 5000 to 100,000 lx. Converted into the luminance value (cd/m^2) closer to visual perception, this means that surfaces can be perceivable in a luminance range from 0.0015 cd/m^2 to almost 1 billion cd/m^2, the luminance of the sun, with perfect dark adaptation.

The amount of light, surface brightness and visual acuity interact in a mutually dependent way (Fig. 2.22).

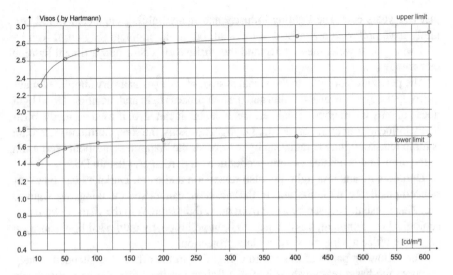

Fig. 2.22 Dependence of visual acuity on luminance, according to Hartmann (1970)

However, the differences in brightness must not be too great in the field of vision at any given time. Then there is the threat of 'dazzling' (Loss of information due to excessive luminance in the visual field) and 'darkness' (invisibility of surfaces in the visual field). When moving in spatial environments with different brightness levels, the visual system needs time to adapt to the situation. This adaptation process of the eye and visual system is called the adaptation process and characterizes the adaptation of the light perception to the surrounding brightness of space and its material surfaces.

In addition to the surrounding amount of light, the colour of the light is fundamental to the quality of perception of a visual environment, which is measured in Kelvin. The quality of colour rendering quality is the most important aspect of light colour.

In the exterior space, the light colour and its change are primarily caused by the position of the sun, and the meteorological conditions are modulated at a given time. Here, too, there are fluid processes of change that are usually not consciously perceived.

The following examples show different colour temperatures. The data looks at different times of the day and meteorological conditions in Central Europe (Fig. 2.23).

- Sunrise 3500 K,
- Morning, overcast sky 5000 K,
- Noon, cloudy sky 6000 K,
- Afternoon, cloudy sky 4200 K,
- Twilight 3500 K,

Fig. 2.23 Prevailing colour temperature over the course of the day

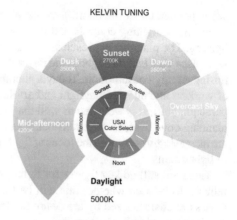

KELVIN TUNING

- Sunset 2700 K.

A humane lighting of the room, oriented to the needs of the human being, consists at least of the factors light quantity and light colour. It can be assumed that active (direct light in the sense of radiation, e.g. solar radiation) and passive (indirect, reflected light from surfaces) light are also components of optimal light.

Contemporary light in office environments consists of the interaction of visual, biological and psychological lighting qualities. Only for the amount of light (illuminance in lux) is there a 'state-of-the-art', laid down in DIN and the Workplace Act. Although there are expert opinions and justifiable assumptions for light colour and the proportion of direct and indirect light, the state of research is not yet universally valid. Only a few years ago, thanks to LEDs, did the technical possibilities of designing the light colour and radiation components of a workplace luminaire in a way that is technically and creatively feasible in today's office environments become available.

Currently, the term 'Human Centric Lighting' (HCL) for lighting design that takes into account not only the direct visual effects but also the non-visual effects of light. The term itself places people at the centre of planning, combined with the demand for an interdisciplinary approach to planning.

2.4.3 The Planning Process

As already described, the light itself is invisible. Perceivable is only the brightness of the self a light source and the light reflected from surfaces. For the light between a light source and the irradiated object 'on the way', we have no sense organ.

Since the end of the 1990s, daylight has generally been calculated using daylight calculation programs when planning office lighting. The reason for this is the increasing use of daylighting software for calculating the amount and distribution of daylight. In the vast majority of cases, however, this has happened and continues to happen under the premise of planning potential savings for artificial lighting by means of lighting controls and the associated reductions in the use of artificial lighting. The planning approach here is ultimately to keep the amount of light at the workplace constant throughout the day with the aim of meeting standards and optimising energy consumption. Today's planning discussions about circadian rhythms of workers and adapted daylight and artificial light inputs represent a fundamental paradigm shift. It is welcome as well that lighting and performance are not the sole planning objectives, but that well-being and willingness to perform at the workplace are increasingly becoming the focus of lighting planning as objectives in the direction of light quality.

There are still no key figures for light quality that have predictive quality, and it remains to be seen whether this will be the case. However, modern planning processes are characterized by the desire to identify and implement different exposure requirements of individual users.

Proven photometric quality criteria for artificial light are still valid and must be observed, these are:

- visual reference to the outside,
- sufficient brightness level of the workplace, immediate workplace environment and room,
- optimal contrast rendition by very good colour rendition of all used artificial light sources,
- authentic body and surface reproduction by means of realistic shading,
- possibility of individualising the light at your own workplace.

All of the above points are factors that are stable over time. These criteria are now supplemented by the requirements for health-promoting light, which is:

- experiencing the variation and dynamics of daylight at the workplace.

In recent years, the terms 'Human Centric Lighting' (HCL) and 'Active Light' established. The aim is to create lighting environments that take into account the interplay of light with the employee's 'inner clock'.

The current basic problem is that the visual and non-visual effects of light do not follow the same laws and functions (Fig. 2.24). The definition of photometric quantities is based on the sensitivity of rods and cones, where the mean value in the green range is 555 nm. The one for the non-visual effect is shifted to the blue range, where the mean value is 480 nm. Both functions overlap only to a lesser extent. This means that the most common lighting parameter used in lighting design, illuminance (lux), is initially not useful for planning non-visual effects of light.

The ratio of these two sensitivity curves is currently the subject of research (Fig. 2.25). An illuminance of 250 'melanopic' lux is considered sufficient for the light to have an activating effect; at a colour temperature of 4000 K (neutral white), this corresponds to a photopic illuminance of 444 lx at the eye. As no measuring instruments for melanopic light are generally available at present, tables can help to determine the melanopic effect of different types of light (cf. Plischke 2015).

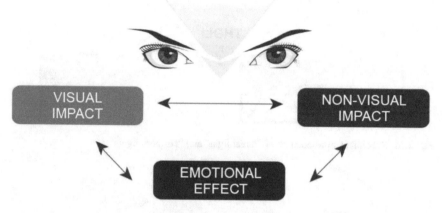

Fig. 2.24 Visual, emotional and non-visual effects of light

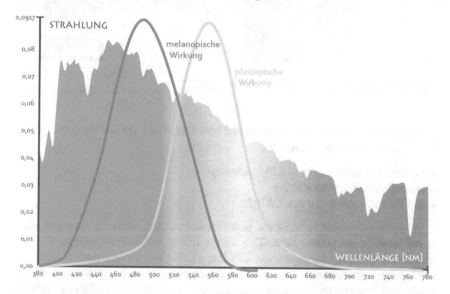

Fig. 2.25 Melanopic and photopic spectrum of action

Analysis of daylight conditions

Without a differentiated analysis of the daylight conditions at the workplaces during the course of the day and generalising over the year, the conception of workplaces with individually experienced quality of stay is not possible.

Electromagnetic radiation, however, is decisive for the conception of rooms and thus for light guidance. Louis Kahn describes these optical arrangements as the 'space in between in which the light is' (Kahn 1991). Louis Kahn's thought processes are

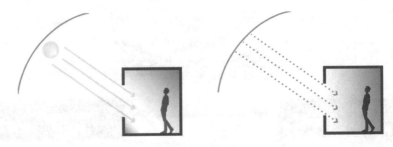

Fig. 2.26 Schematic representation of "direct light" and "northern lights"

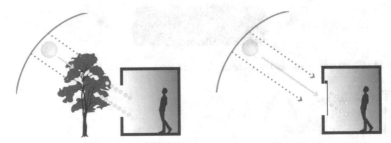

Fig. 2.27 Schematic representation of "refracted light"

extremely practical in the description and conception of light and light guidance in rooms. He distinguishes between

- the direct radiation light of a light source,
- the undirected northern lights,
- and the refracted or modulated light.

As soon as people consciously perceive objects and surfaces, they can speak of refracted or modulated light. This means that if undirected daylight, i.e. light without direct sunlight components, falls into a living room with a normal-sized side window, the room appears to conform to expectations in such a way that people do not consciously evaluate the light, they see the room surfaces and the objects in the room. The light is unweighted, i.e. unconscious.

If the light situation changes and direct sunlight additionally penetrates through this window, the window geometry will appear on the floor and walls of the room. The sunlight is 'modulated' and leads to a conscious perception of light and the subjective evaluation 'that beautiful weather is outside'. At the same time, an external cover is created.

The smaller the window in this room, the greater the psychological significance of this daylight opening. The quantity of light and the psychological significance of light in space are not interdependent. This applies to both daylight and artificial light situations. Planning here is not easy and never intuitive (Figs. 2.26 and 2.27).

Fig. 2.28 Incidence of light in a conventional side window

It is possible to make interiors appear only bright. Here, light dominates the impression of the room. Room surfaces become reflective surfaces. The subjective meaning of furnishings is limited to their functionality.

These interactions are illustrated using a project example in Sect. 3.3 is illustrated.

As shown in the first few paragraphs, man and his visual system have been able to be developed. Daylight has a different effect than artificial light. The following definitions are

- the amount of light generated by daylight at the workplace near the window is much higher than the standard-compliant values of artificial light at the same location,
- it has significantly more short-wave, activating radiation components,
- daylight-oriented work zones can usually be furnished more freely,
- window-related workplaces are generally more circadian effective and therefore have more significant timing incentives than window-related workplaces.

The effective use of daylight should therefore be the basis for circadian lighting in the workplace.

In conventional office rooms with side windows, the room zone is The width of the roof, in which sufficient daylight is available for office work, must not exceed 4–5 m (Fig. 2.28).

The right artificial light at the office workplace

At present, almost every well-known luminaire manufacturer has developed lighting systems with daylight-dependent light quantity control and light colour change, thus expanding its portfolio to include luminaires with biodynamic light management functions. Whether these lights then produce the desired psycho-/physiological effects, however, can often only be assumed in the concrete case.

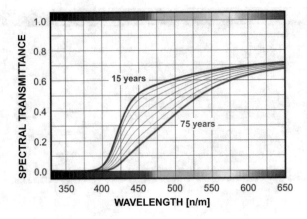

Fig. 2.29 Transmission of the eye lens in different age groups

Contemporary office lighting should consider the following parameters

- Light quantity/light intensity,
- Light direction,
- Light colour,
- Variability of light over time and its course.

According to current knowledge, the average illuminance values of 500 lx on the usable area and 300 lx in the immediate surrounding area for artificial lighting planning in the office, which are determined by standards, are in any case insufficient in terms of quantity for serious human centric lighting.

The 1975 lighting manual already stated that older people (60 years) need more light than young people (20 years). However, they can perform a given visual task at 900 lx just as well as 20-year-olds. It is explicitly stated that 'a high lighting level can therefore create equal working conditions for all workers' (Spieser und Herbst 1975, p. 37). The physiological development of the eye is influenced by an increasing need for light with age. is marked. The lens of the eye and the cornea are subject to an ageing process. There is a decrease in contrast on the retina. This process can be highly compensated by more light.

Figure 2.29 shows the decrease of the spectral transmittance of the eye. In particular, the degree of transmission for blue light decreases continuously with increasing age. As a result, the perceived activating portions are also reduced.

The most important basic lighting parameter in lighting design at present is illuminance. The current standard value for office work is 500 lx at the desk level.

Even in the 1970s, 300 lx was still common in the office. Today, the valid standard value for office work according to EN 12464 is 500 lx in the field of activity. and 300 lx in the immediate surrounding area, see Fig. 2.30. A minimum of 200 lx is required for vertically oriented visual tasks.

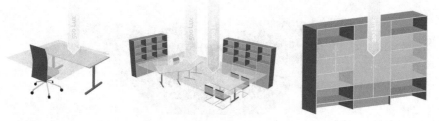

Fig. 2.30 Illuminance in the office according to DIN 12464

Fig. 2.31 Luminance
sensitivity range of the eye

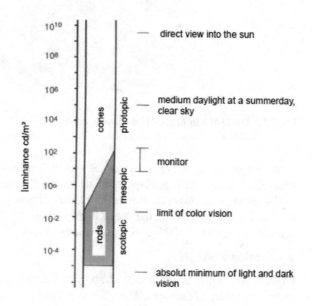

The illuminance, the product of the amount of light (lumen) and the area (sqm), is not visible to the eye itself, but the presence of a sufficient amount of radiation is a necessary condition for brightness perception.

Although illuminance is not visible, the 500 lx required for office work is not chosen arbitrarily or the result of ignorance of psychological perceptions (Fig. 2.31).

According to Fig. 2.31, with an ambient illuminance of 500 lx, full colourfulness is achieved in healthy eyes. The necessary visual acuity for reading texts and writing is also achieved at the given luminance of approx. 150 cd/m^2 on white paper with black writing, provided that the persons are normal and middle-aged. The '500 lx' has also always been a compromise between technical feasibility, avoidance of glare and, in particular, energy input in workspaces. It is not the ideal value for ideal illumination and the associated visual brightness at the workplace. In conventional lighting systems with fluorescent lamps, approx. 4 W/m^2 per 100 lx are introduced into the room.

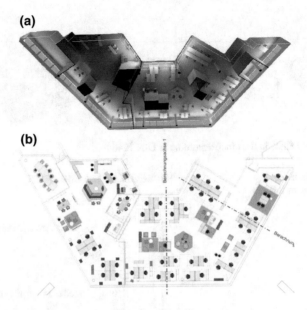

Fig. 2.32 Daylight distribution in the example office

With energy inputs of over 20 W/m², it was only possible with disproportionately great effort to condition workspaces using air-conditioning technology.

This is one of the biggest advantages of LED technology. With a specific connected load of 4 W/m² per 100 lx, the use of highly efficient LED solutions today enables illuminance levels of 1000–1200 lx on work surfaces.

Application Example

Given is an open-plan office with a trapezoidal floor plan. The windows are located on the long sides of the room. The office workplaces are oriented to the windows. In the central zone of the room, there are work zones that are used temporarily, including co-working and meeting zones for spontaneous meetings between employees. There are also two meeting rooms in this room zone. These are daylight-oriented with side windows, but spatially separated from the open-plan office.

Figure 2.32 shows the daylight distribution in the open-plan office. The sizes of the window areas are normal and correspond to DIN 5034 'Daylight indoors'. Nevertheless, in relation to the other closed room surfaces of the room, the perceived window surfaces in the visual field are rather small. Here, there is the danger of adaptation glare. It is absolutely necessary to place a glare shield in front of the windows on the inside in order to adjust the window luminance to the interior luminance.

Only daylight, i.e. without additional workplace lighting, supplies only the workplaces directly oriented towards the window. Even from a room depth of 4 m, daylight will have to be supported by artificial light at all times.

The first planning approach was to place as many workplaces as possible in the immediate vicinity of windows.

A free-standing luminaire with separately switchable and dimmable direct and indirect components was selected as the lighting system.

Core working time starts at 7:00 and ends at 18:00. At 7:00 a.m., the artificial lighting is automatically switched on. The indirect component of the free-standing luminaire and a bright white ceiling achieve an average illuminance of 300 lx across the entire area. This value of 300 lx is never undercut at any time during core working hours.

If the employee is present at his workplace, it is at his discretion to put the direct share into operation. This direct component can reach 500 lx on the entire work surface.

Each floor lamp is equipped with brightness controls, and all floor lamps communicate wirelessly with each other in this room.

The colour temperature for indirect and direct components in normal operation is 4000 K, neutral white. The colour rendering (CRI) is above 90.

Biodynamic light is used on two periods of the day—between 08:00 and 10:00 in the morning and between 13:00 and 14:00 in the afternoon. In these phases, the light colour of the indirect component changes to a colour temperature of 5500 K, daylight white. The illuminance of the indirect lighting is increased by 200 lx. After 14:00, the colour temperature returns to 4000 K and the illuminance of the general lighting returns to the initial value of 300 lx (Fig. 2.33).

All light changes must take place slowly and continuously. Ideally, the regulation should be imperceptible and without conscious attention on the part of the employees.

The co-working and meeting zones in the centre of the open-plan office are differentiated with light for specific uses. This applies both to the choice of luminaires and their light distributions. In the meeting zones, the choice of luminaires depends on the arrangement of the furniture, the light colour depends on the surface colours of the materials used.

A special feature in the core zone is a backlit light wall, in this case, printed with a natural motif. This lighting element is integrated into the light cycle of the free-standing luminaires at the workplaces and helps to maintain the adaptability of our visual system as stimulating light or substitute brightness in the absence of daylight (Fig. 2.34).

2.5 Body—Food and Mood

'One shall offer the body something good, so that the soul may desire to dwell in it'. (Teresa of Ávila). A person with a healthy body feels well, has a better mood and can therefore also work happier and better. A healthy workplace can contribute a lot to this, because only a healthy employee is motivated and performs well. Many companies now provide water and hot drinks, such as coffee and tea and fruit, mostly

TASK AREA LIGHTING WITH BIOLOGICAL EFFECT

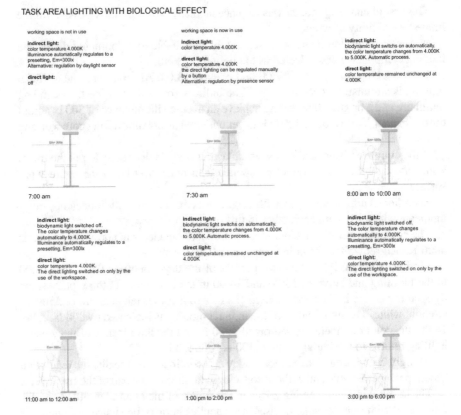

Fig. 2.33 Biodynamically activating light in the course of the day

in the form of apples, is also available. But what can companies do to encourage employees to be health-conscious about their own bodies?

2.5.1 Healthy Food

Health-conscious nutrition has now become a topic of social interest and is, therefore, a popular topic of discussion not only among friends, but also in the tea or coffee kitchen. Working conditions have changed in many areas. The employees in the office do not move much and eat 'conveniently', often with too much fat and sugar. Or they still feed like the generations before us, who consumed much more energy a day. The result is that half the Germans are overweight. One in five is obese, has cardiovascular problems and is at risk of developing diabetes. Occupational and private demands are increasing and with them a permanent stress burden that can lead to serious health

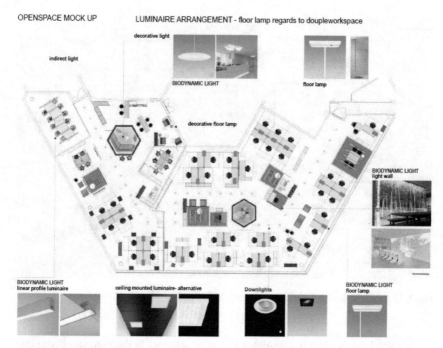

Fig. 2.34 Selection of luminaires for the entire open-plan office

consequences—from exhaustion to burnout. A healthy diet can make a contribution to counteracting this.

Most adults spend at least eight hours a day at their workplace during the week. People eat in the breaks, mostly hectically and often also indiscriminately. Especially people who work shifts eat irregularly and often eat an unhealthy diet. Companies could help their employees and support them with healthy ideas. For example, you can initiate events with a nutritionist, have healthy food delivered or go a step further and enjoy healthy and varied food in the canteen.

A number of start-ups are breaking new ground in their attempts to offer fast and healthy meals in everyday office life. They are called for example littlelunch.de or EatFirst. Everyone should have the opportunity to get fresh and healthy food during their lunch break. Healthy nutrition and convenience should be brought into harmony, without cooking, stress and junk food. Some suppliers still have to heat the dishes, others work together with restaurants and deliver the dishes ready to eat.

Start-ups have thus created a new segment: Not because it is particularly creative to deliver food, but because the focus has shifted. In times when conscious nutrition is a big issue, but stress and strain on the job are constantly growing, these start-ups have recognised a gap. Although this principle is still in its infancy in Germany, the delivery areas covered extend mainly to the large cities and are therefore still very limited in some cases. But it is an interesting market field and will certainly continue to develop.

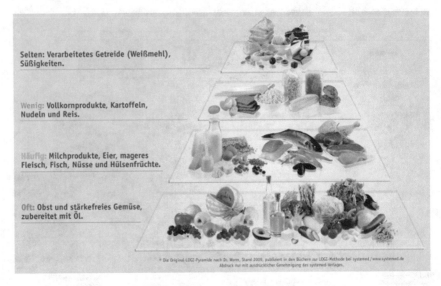

Selten: Verarbeitetes Getreide (Weißmehl),
Süßigkeiten.

Wenig: Vollkornprodukte, Kartoffeln,
Nudeln und Reis.

Häufig: Milchprodukte, Eier, mageres
Fleisch, Fisch, Nüsse und Hülsenfrüchte.

Oft: Obst und stärkefreies Gemüse,
zubereitet mit Öl.

* Die Original-LOGI-Pyramide nach Dr. Worm, Stand 2009, publiziert in den Büchern zur LOGI-Methode bei systemed / www.systemed.de
Abdruck nur mit ausdrücklicher Genehmigung des systemed-Verlages.

Fig. 2.35 LOGI Nutrition, according to Worm (2015), p. 59

Healthy food is already somewhat better established in the canteen. Some canteen operators already use the LOGI method for cooking. LOGI stands for 'Low Glycemic and Insulinemic Diet', which can be translated in German as 'Ernährungssungsmethode zur Förderung eines niedrigen Blutzucker- und Insulinwertes'. This reduces blood fat levels and effectively prevents the formation of fat deposits and eliminates cravings and weight loss. The method is based on the findings of Prof. Dr. med. David Ludwig, endocrinologist at Harvard University Hospital in Boston (USA). In a technical paper, he presented an alternative nutrition pyramid (Fig. 2.35), which was then disseminated by Dr. Nicolai Worm in the German-speaking world (Worm 2015; http://www.logi-aktuell.de/logi-methode/logi-pyramide).

The basis of LOGI nutrition contains vegetables, mushrooms, salads and (low-sugar) fruit. They have a low blood sugar effect, satiate well due to their high volume and fibre content and at the same time ran a wealth of vitamins, minerals and secondary plant substances. Healthy fats are also at level 1, because they provide important nutrients (essential fatty acids and fat-soluble vitamins) and also contribute to satiation and also provide better blood fat values—and a lot of taste. High-quality fats are olive oil, rapeseed oil, walnut oil, linseed oil and butter.

On the next level, there are protein-containing foods, which also have a low blood sugar effect and long satiety. Thus, the basic foods can be supplemented with meat, fish, eggs, milk and milk products or pulses and nuts—gladly with every meal.

On the last two levels of the pyramid, there are foods with a strong or very strong effect on blood sugar, which should be used sparingly.

A very good study, not influenced by the food industry, confirmed these principles for a healthy diet in mid-2017 (cf. Dehghan et al. 2017).

It does not have to be LOGI food in the canteen, but you can contribute a lot to a healthier diet with more conscious offers. At Diamant Software in Bielefeld, for example, there are two meals cooked on site, which pay attention to a reasonable diet. The company pays the kitchen staff, who come from a neighbouring restaurant, but cook locally in the company for about 100 employees. The employee pays the 'material costs', which corresponds to a cost contribution between 3 and 4 euros.

2.5.2 Food as a Community

Companies that have recognised that eating always has something to do with a sense of community and exchange go one step further. This is why in Scandinavia, for example, there is in many places in the company an area with large tables as well as an outdoor area, which can be used at lunchtime for a communal meal. It is used as a meeting zone during the rest of the time. Either cooking is done on site or a delivery service supplies the company with freshly prepared food, which is consumed by all together.

The project "The Kitchen" by Studio Olafur Eliasson in Berlin in a converted brewery at Prenzlauer Berg goes even further. During the week, the studio team—artists, architects, craftsmen, technicians, art historians and archivists—gather around the tables for lunch. Visitors are also cordially invited. In the past, everyone had to leave the office at noon if they wanted to eat something sensible. Only the change to healthy, tasty food now gathers everyone around one table. At that table, there are encounters and unplanned but desired exchanges between all employees in the office. Olafur recognised the potential, eating together creates a connection between all and ultimately serves mutual inspiration. For Olafur Eliasson, cooking means "caring for others. It is a gesture as generous as it is hospitable, strengthens social ties and is an expression of ideas that are not only about food itself, but also about giving and sharing" (Eliasson 2016, p. 12).

It is certainly not necessary to take every lunch together in a larger company, but you can consider taking a healthy meal together 1–2 times a month.

Companies such as Google and Co. also take advantage of this principle of eating. They provide their employees with all meals, snacks and drinks and thus achieve a high level of attendance and a longer stay at the workplace.

2.5.3 Activity-Based Working

In addition to a healthy diet, it is also about exercise in the workplace and how to promote it. Today the knowledge worker spends a lot of time in a sitting position at the desk or in other places. In addition, there is often an unhealthy posture in front of the computer, which allows little change. Many companies now support standing workplaces and often advise their employees to change positions.

'Many managers know that workers perform better, are proactive and exchange ideas in an environment of interaction and communication with colleagues' (Meakins 2016). Microsoft has changed their office concept company-wide based on the concept of activity-based working (ABW) changed.

Microsoft's goal was a radical transformation of the work environment. The company wanted to become a living example of progressive and productive jobs. Steven Miller, Business Group Director of the Microsoft Office Division, explains in an interview: "The fact is that people work with different devices in different ways. We wanted to give our employees the opportunity to be more flexible in a performance-enhancing environment. Activity-related work helps us to do this because it blurres the boarders between departments, typical of large companies". (ibid.)

In order to successfully implement ABW, various things have to work well together. Trust comes first. This concept, which is based on an optimal corporate culture, works not without mutual trust. . You need motivated employees and a well-thought-out concept and, of course, the right technology.

ABW means in its implementation to provide employees with changing working environments, which offer the right and supportive environment depending on the task. Inflexible room systems dissolve the requirements depend on the respective activities. Everyone chooses the right environment for the task and his needs, which he can change again several times during the day. This can be done in a wide variety of areas within the company, such as central zones, communal areas, work cafés, project and conference zones, retreat areas and team areas.

In order to further support the health aspects, smaller sporting activities can also be incorporated. For example, there are now treadmills on which you can work at the computer in a slower pace, movable standing cushions to enable dynamic standing, trampoline , in order to relax or completely different seating and standing furniture, on which one can balance and swing and thus relieve the spinal column.

The core of the ABW is to give the employee the opportunity to choose where and in which environment he wants to work and to change his position several times a day, i.e. to alternate between sitting, standing and moving.

2.5.4 Hygiene in the Workplace

Wherever many people meet on a daily basis, viruses and bacteria can be found. This is especially true for an office where workplaces are changed. Therefore, special emphasis should be placed on hygiene in the workplace. Simple measures are already sufficient to ensure more cleanliness and thus more health at the workplace.

General office rules should also cover the area of hygiene for the benefit of all. The following guidelines are examples of how they could be defined:

- Once a week, clean the keyboard, telephone receiver and computer mouse with a fat-soluble detergent (in many companies these things are also personally assigned to an employee).

- Air the work areas several times a day at fixed times to reduce the number of viruses and bacteria in the room by exchanging air.
- Wipe the refrigerator regularly with hot water with cleaning additive.
- Dispose of expired food.
- Wash hands regularly and provide disinfectants in washrooms (for non-medical staff: hand disinfection is not necessary after thorough hand washing; in certain situations, it can replace hand washing—for example, in situations where going to the sink is perceived as being too time-consuming).

It has to be clear to everyone: Hygiene is a task for everyone, beyond responsibilities. Regular clean-up rounds for everyone in a playful way can help. Ultimately, a feeling must arise that says: 'This is our office—and we are all responsible for the cleanliness here'. This means that all colleagues are equally required to pay attention to hygiene at the workplace, and everyone benefits from it at the end of the day.

2.5.5 Rules (Room Use/Handling with Colleagues)

There is a lot of talk about corporate identity and also about image maintenance and corporate design. In all efforts to achieve a positive image, it should not be forgotten: Charisma takes place from the inside out and the attitude of the employees is the most valuable capital. Polite behaviour in dealing with customers can only have an authentic effect on them if there is a cooperative tone behind the scenes (see Sect. 10.4 Dealing with colleagues) (rules).

A lack of cooperation, consideration and foresight leads to ineffective internal communication, information deficits, loss of motivation, mistakes and losses. Positive internal communication, on the other hand, creates a pleasant working atmosphere, promotes team spirit and forms a solid basis for operational and human success.

In the end, however, it is also a matter of agreeing on certain rules among themselves, how to deal with each other and how to use the different rooms in such a way that the others do not feel disturbed and everyone has the right and opportunity to use the different areas.

The rules, how one behaves at the workplace and which rooms one uses for what, are very individual for a company. The best way to create them is to work together on a department-specific basis in a workshop, and often a change process is an ideal solution. The challenges and fears of the employees can be very well reformulated into positive sentences and thus rules can be set in a playful way. A funny and good example is Microsoft's prohibition sign in the Netherlands for the permanent use of retreat areas: 'Camping prohibited' (Fig. 2.36).

If someone does not obey the rule, you simply put this sign up for him without comment, smiling. No one has to scold or justify themselves and long discussions or even a quarrel are avoided.

Fig. 2.36 Camping forbidden (Microsoft, Amsterdam, photo Christine Kohlert)

In a game rules workshop, rules for the behaviour and use of the areas are jointly agreed upon. This can result in a poster that can be hung up in a clearly visible position. After some time, another workshop will be organised to review the rules and, if necessary, supplement and discuss what makes sense and is important. Such rules relate, for example, to the use of headsets, the volume during telephone calls and conversations, but also the duration of the use of retreat rooms (see also checklist, Sect. 10.1).

After all, an office is a community of people and, as in a private environment, you have to spend a certain amount of time with different people every day. This is better done if you get involved with others and try to understand and accept other opinions. It is not good for you always insisting on your own rights, you must openly address problems or disagreements and search for common solutions that are acceptable to all.

Literature

ASR A3.6: Technical rules for workplaces, ventilation, January 2012.
BGI 7003: Assessment of indoor climate, Berufsgenossenschaftliche Information für Sicherheit und Gesundheit bei der Arbeit, 2008.

BGI 7004: Climate in the office—answers to the most frequently asked questions, January 2007.

Dehghan, M., et al. (2017). Associations of fats and carbohydrate intake with cardiovascular disease and mortality in 18 countries from five continents (PURE): A prospective cohort study. *The Lancet*, 390/10107, pp. 2050–2062. Available at: http://www.thelancet.com/journals/lancet/article/PIIS0140-6736(17)32252-3/abstract. Access on October 5, 2017.

DIN ISO 16000-28: Indoor air pollutants—Part 28: Determination of odour emissions from construction products using an emission test chamber, December 2012.

DIN EN 7730 (200G): Ergonomics of the thermal environment—Analytical determination and interpretation of thermal comfort by calculation of PMV and PPD indices and local thermal comfort criteria.

Eliasson, O. (2016). *The kitchen*. Knesebeck: Phaidon Press.

EN 13779: General principles and requirements for ventilation, air conditioning and refrigeration systems.

EN 15251: Indoor climate input parameters for the design and assessment of energy performance of buildings—indoor air quality, temperature, light and acoustics, May 2007.

EN 16798-1 Draft 9: Guideline for using indoor environmental input parameters for design and assessment of energy performance of buildings, May 2016 (draft).

FGK Status Report 8, Questions and answers on indoor humidity, C. Händel, November 2007.

FGK Status-Report 17, Evaluation of indoor climate, Fachverband Gebäude-Klima, R. Hellwig und C. Handel, May 2012.

Hartmann, E. (1970). *Lighting and vision at the workplace*. Munich: Goldman.

Health assessment of carbon dioxide in indoor air, published by the Federal Environment Agency in the Bundesgesundheitsblatt -Gesundheitsforschung - Gesundheitsschutz 2008, pp. 1358–1369.

Hugentobler, W. (2016). Indoor air humidity under the influence of increasing energy efficiency and socio-economic changes, TGA Congress Berlin.

http://www.logi-aktuell.de/logi-methode/logi-pyramide. Access on October 5, 2017.

Kahn, L. I. (1991). *In the realm of architecture*. New York: Rizzoli.

Kempski, D. v., Ziegler, H. (2000). Addition of room air essences and room air quality. *Heating, Ventilation and Building Services*, 51/2, 68–73.

Meakins, M. (2016). Activity based working—From the office to the activity-based workplace. http://www.think-progress.com/de/effizienz-am-arbeitsplatz/activity-based-working-vom-buro-zum-aktivitatsbezogenen-arbeitsplatz/. Accessed on June 15, 2016.

Plischke, I. (2015). The transformation. German short television film, broadcast on 18.11.2015 on Bayerischer Rundfunk.

Schmitz, K. W. (2014). *The strategy of the 5 senses*. Weinheim: Wiley.

Scofield, C. M., & Sterling, E. M. (1992). Dry climate evaporative cooling with refrigeration backup. *ASHRAE Journal*, 49–54.

Seifert, B. (1999). Indoor air guidelines—The assessment of indoor air quality using the sum of volatile organic compounds (TVOC value), Umweltbundesamt im Bundesgesundheitsblatt - Gesundheitsforschung - Gesundheitsschutz 1999, pp. 270–278.

Spath, D., Bauer, W., & Braun, M. (2011). *Healthy and successful office work*. Berlin: Erich Schmidt.

Spieser, R., & Herbst, C.-H. (1975). *Manual for lighting*. Food: Girardet.

VDI 3804, Ventilation technology for office buildings, March 2009.

VDI 3882, Part 1 and 2: Olfactometry—Determination of odour intensity (Part 1), Olfactometry—Determination of hedonic odour effect (Part 2), October 2008.

Workplace Ordinance ArbStättV, August 2004.

Worm, N. (2015): LOGI method. Happy and slim. 13th, overworked, updated. Edition. Lünen: Systemed Verlag.

Chapter 3
Colour in Theory and Practice

Werner Seiferlein, Rudolf Kötter⑩ and Katrin Trautwein

3.1 Introduction to the Chapter

Colour and lighting interact with each other. The colourfulness of a room has many effects on the employees. Different lighting and colours also demonstrably improve motivation, concentration and performance.

However, the question arises whether all people see colourfully in the same sense. There is an interaction between language and perception.

The following striking statements on this topic are known from the literature and should be listed here, namely:

- that the employee is in a "blue" environment can concentrate more on the task—but the red environment tends to turn attention away from the task (see Stone and English 1998).
- that colours have an influence on the cold/warm sensation in a blue-green painted room one freezes already at 15 °C, in an orange one however only starting from 12 °C; this is reflected also in the vernacular as the adjectives "ice blue" and "fire red" impressively confirm.
- that warm colours are often associated with interest, communication and love, but also with power and aggression—to the observer they are active, exciting and invigorating, whereas cold colours convey distance, coolness an impression of

W. Seiferlein
Technology Innovation Management, Frankfurt/Main, Germany
e-mail: werner.seiferlein@timoffice.de

R. Kötter (✉)
FAU, Erlangen-Nuremberg, Erlangen, Germany
e-mail: rudolf.koetter@fau.de

K. Trautwein
kt.COLOR AG, Uster, Switzerland
e-mail: trautwein@ktcolor.ch

© Springer Nature Switzerland AG 2020
W. Seiferlein and C. Kohlert (eds.), *The Networked Health-Relevant Factors for Office Buildings*, https://doi.org/10.1007/978-3-030-22022-8_3

thinking, freshness and reason; the effect of cold colours is often described as passive, calming, relaxing and refreshing (cf. Nüchterlien und Richter 2008).

- that warm colours should be preferred for furniture, walls and floors and that dark, oppressive colours such as black, brown and grey should generally be banned from the office.

Is it as easy to determine the colour as it is shown above? In order to understand the meaning and the goal to be achieved in dealing with colours, one has to go deeper into the subject and research it.

What can be determined from the selection of the relation to the ability and effect of the colours? The result that formulates derivations for the future (practice) from the origin (theory) will be exciting.

Is there perhaps a need for networking, in which the isolated individual statements can be combined to form a colour concept?

Does the selection of colours for walls, ceilings, floors, furniture, etc. enjoy the same importance as the planning of an air conditioning system? Specialists are usually involved in the planning of an air conditioning system. For the selection of colours, without reflecting on colour concepts, it is usually the hierarchically superior employees who feel called upon to carry out the colour selection.

For this reason, this section will examine the subject of "color" more deeply and thoroughly. This puts the importance of the subject of "colour" in the right light.

In order to do justice to the goal, the subject of colour is treated both philosophically, i.e. theoretically, and operatively, i.e. practically. Greats such as Newton, Goethe and others have dealt with the subject of colours, from which we should profit today. We will see that in some cases, there are interfaces that develop similarities between theory and practice. The resulting interface is to be shown as an example for a point of view:

- **Theory (Kötter, Sect.** 3.2): "The rose is red by day and by night", but the ambient light determines the colour visible to man.
- **Practice (Trautwein, Sect.** 3.3): As we have seen, colour is the interaction between light and surfaces that makes the world visible to us thanks to our constant absorption of light energy.

Interfaces of this kind are often to be recognized in the comparison of theory and practice.

With this topic, it is clear that without specialists the planning would take place just as unprofessionally as with a ventilation system.

3.2   The World in Colour as a Challenge for Philosophy and Science

3.2.1   Introductory Remarks

The phrase "This rose is red" has always been regarded as a prime example of a statement that is simple in every respect. If the marking is resolved logically, its simple explanatory structure can also be seen: "This object is a rose and this object is red". One convinces oneself of its correctness or falsity by simply looking at it (red) and recalling some features that can also be grasped by simply looking at it, which determine the use of the predictor "rose".

Since antiquity, we have had the notion that our knowledge of the world must, in the final analysis, be built up from such simple sentences that can be verified at any time with regard to their validity. The classical empiricists of the European Enlightenment postulated that every experience consists exclusively of impressions that we get from the world beyond us through our senses. Of course, we know that our senses can occasionally deceive us, but we also know that these deceptions can in turn be exposed, corrected and avoided through the skilful use of our senses. The logical empiricists of the twentieth century then added a linguistic counterpart to this sensual foundation of experience: The protocol, which linguistically fixes the subjective sensual impression, is considered fundamental in the sense that it can neither be derived from theoretical considerations nor corrected by them [important texts on the protocol debate can be found in Schleichert (1975) or Damböck (2013)].

In other words, the special feature of empirical knowledge can be seen in the terminology and form of the sentences with which it is expressed. This conviction has been summed up in a criterion of meaning:

1. Every meaningful term either refers itself to directly observable objects or can be expressed with the help of such terms.
2. In the final instance, the justification of an empirical assertion is always made with reference to the results of direct observation.

I see here and now this red rose (protocol record) and may therefore claim: "This rose is red". Ultimately, all empirical judgements should be based on this pattern. But even this sentence, so harmless at first sight, has a lot to offer. Especially statements about colour phenomena can make it clear that the hope of arriving at safe and reliable empirical findings by simple and straightforward means is in vain. "Colours encourage us to philosophize", Ludwig Wittgenstein said (Wittgenstein 1984, p. 544). And he was undoubtedly right, as we will see below.

3.2.2   Colours as Properties of Things

The transition from subjective protocol to objective empirical judgement only works with the aid of a metaphysical premise. There is a real world outside of us, and its nature does not depend on whether we look at it or not. Even if the premise sounds rather harmless, it is of fundamental importance, since only under its assumption is the step from subjective experience to objective experience possible.

Suppose it was night and I wanted to go to bed. After I have extinguished the light in my living room and made my way to the bedroom, I remember that I absolutely have to give the roses on the living room table fresh water. In accordance with the above premise, I can of course assume that they will still be standing in their place in their vase. But are they still red? That is, does a rose retain its colour in the dark and can it only not be seen because of the darkness, just as one cannot see (but feel well) its species-determining characteristics such as flower structure, leaves and spines in the dark?

There are philosophers from the camp of analytical philosophy who maintain a metaphysical physicalism as a worldview and claim exactly that. Your position is described with "color realism"or "colour objectivism" [for the status of the discussion see Thompson (1995) and Dorsch (2009)]. In this sense, D. M. Armstrong, for example, states that the colour f of an object (1) exists regardless of whether it is perceived and (2) is nothing other than the bundle of light waves reflected from its surface ("reflectance"). Under *normal* circumstances, this bundle of light waves is then perceived as colour f by a *normal-sighted* observer (Armstrong 1969, p. 119 f.; somewhat refined Byrne and Hilbert 1997).

Now we know that the "red" of the rose comes from the dye cyanine, which is stored in the cells of its petals. This dye is undoubtedly retained in the rose, even if it becomes pitch-black. Chemists could provide proof of this, if necessary even in the dark. But cyanine is characterized as a chemical compound by its molecular structure, and "color" is not a chemical structural element. The fact that cyanine appears red is a fact that only presents itself to the viewer when viewed in light. The reductionist idea of trying to equate "being red" with "possessing cyanine" in the case of the rose thus does not take us any further. In addition, there are cases where, on the one hand, objects with different reflectance in the physical sense convey the same colour impression to the observer [the so-called metamere, cf. Hardin (1986) and (2014)] and, on the other hand, people of different ethnic origins combine the same reflectance with a different colour impression. And finally, the combination of colour impressions with the physical properties of surfaces proves to be unfortunate for the treatment of phenomena such as the rainbow, the northern lights or the coloured shadows on snow, all of which cannot be explained by "reflectance" in the sense explained above. The colour lens specialist meets these objections by referring to the "normal circumstances" and the "normal-sighted observer". But since it is not the philosopher's competence to determine what "normal circumstances" and "normal-sighted viewers" are, the colour objectivist implicitly confesses his lack of competence with his definition: colour is not an issue for philosophy, others have to take care of it.

From the point of view of colloquial language, the speech of "invisible colours" seems as absurd as that of "inaudible sounds". This is not because the natural sciences have not yet succeeded in discovering such phenomena. The reason for this lies neither in the sciences nor in a theory of colour, but rather in what Wittgenstein calls the logic of word use: With colour predicators we designate properties which convey a certain sensory impression, and this only occurs in light. Thus colour words are determined in their linguistic-pragmatic function, and therefore, one is tempted to capture them apodictically: where there's no light, there's no paint. But be careful: the rose is not so completely cut off from the colour in the dark. Although we ascribe to the rose only in light the actual, i.e. directly controllable quality "red" by sensory perception, in darkness we at least ascribe to it the dispositional quality of being *able to appear* red again in light. Depending on the external circumstances, colour predicators therefore have an updated content once, another time a discretionary content. With this thought, Jackson and Pargetter, for example, try to save colour objectivism (cf. Jackson and Pargetter 1997, p. 77).

3.2.3 Colours as Dispositions of Things

Dispositional characteristics are nothing special for us, the predicators by which they are called are often characterized in German by the syllables "-lich", "-bar" or "-keit". Thus the disposition of a glass vase to break when it hits the ground is expressed by the sentence "this vase is fragile"; accordingly I say of a piece of sugar cube that it is soluble in water. A copper cable has the disposition to conduct current when voltage is applied, for which we do not have our own disposition printout. There is no "Stromleitlich" in German, we have to express the disposition somewhat more laboriously; and we also have no word of our own for the fact that the rose in the dark has the disposition to appear red in the light.

Obviously, states in which an object can be assigned an actual property correspond, at least occasionally, with other states in which exactly this property is only present dispositionally. According to the logical empiricists programme, a property is only "really present" if it can be updated. In this sense, disposition expressions would only be empirically meaningful if they could be defined by means of predicators that designate actual properties. There has been no lack of attempts to comply with this demand, but with no convincing result. I suppose most readers would find the following suggestion plausible:

> x has the disposition 'red' in the dark exactly when the following applies: If x were brought into the light, x would appear red.

Here, however, the subjunctive "would" emerges and this means that one enters the field of the so-called counterfactual implications ("counterfactual conditionals"), which is interspersed with so many logical and linguistic-philosophical pitfalls that one is well advised to retreat immediately.

Even the logical empircists, above all Rudolf Carnap as one of their main representatives, had to admit that with their great programme they were stuck with the simple expressions of disposition so widespread in the language. Of course, this is why philosophers have not abandoned the analysis of dispositional expressions. To this day, they have remained test pieces for any epistemological or scientific-theoretical design. But so far, there are no really convincing suggestions as to how dispositional expressions can be integrated into the linguistic-philosophical inventory. A volume has recently been published which gives an overview of the current state of the discussion (cf. Vetter and Schmid 2014). After reading, it must seem incomprehensible that even small children know how to use disposition expressions without difficulty.

One reason for the unsatisfactory complexity of the discussion could be that philosophers have been seduced by language: A disposition expression stands in the sentence in place of the predicate and suggests that a *property of* the object is designated by it. Thus, on the ontological level, one sets out to postulate properties which, although they cannot be shown to be actual, nevertheless exist in some way and are mysteriously and firmly connected both with actual properties of the object and with certain states of the world. This leads to an "ontology of indeterminate possibilities", as I would like to call it: each actual feature of an object was dispositional under different circumstances and earlier times and at the same time contains the disposition to an indeterminate number of possible future actualizations. Thus the glass vase on the table is the actualization of a disposition that contained a heap of quartz sand before melting and forming, at the same time carrying the disposition to break or melt within itself. In every real moment, all possible world processes are arranged dispositionally and which of the dispositions are realized is determined by natural laws with their initial and boundary conditions. One can believe this, but one should not claim to have thrown an illuminating light on our use of dispositional expressions.

If one lets go of the assumption that the structures of the world are contained in the structures of language, then one can open oneself to the thesis that no ontological fact is established with a disposition judgement, but that a complex linguistic-pragmatic fact is expressed. How would one convey the use of an expression of disposition in everyday life? Suppose I said, "This glass vase is fragile". Then, in defence of this claim:

1. This vase *is* actually made of glass.
2. I and many others have *experienced* in the past that objects made of glass are broken when they have fallen from a sufficient height onto a hard surface.
3. I and many others *expect* that this vase will break if it falls from sufficient height onto a hard surface due to the 1st and 2nd.

Or, to return to the example from above: I maintain in the dark living room that the roses have the disposition "red", and by this I mean that they are actually roses that have always shown themselves red in the light so far, which is why it is to be expected that they will do so again when the light is switched on.

Seen from a pragmatic point of view, a dispositional expression is an ingeniously efficient way of bringing together current *findings, experiences* and *expectations* in a single word. If someone wants to doubt the claim of the fragility of the vase, there are exactly three counterarguments available to him:

1. This vase *is not* made of glass, but of plastic.

    Or:

2. The experiences with such vases are *not clear*: sometimes such objects are broken, sometimes not. No concrete expectations can be supported on the basis of such conflicting experiences.

    Or:

3. Only two cases have been observed in which such vases have broken, this experience base is *quantitatively insufficient* to justify a general expectation.

The examples show that the experiences one has to fall back on when using disposition expressions are different. It may be an experience with the object itself (as in the case of the rose or the copper wire), but it may also be an experience with other objects, which in nature are the same as the object in question (as in the case of the glass vase or the piece of sugar). Certainly, the justification strategy to be followed to justify a *statement of* "x is p" is different from the justification strategies to be followed to identify *experiences as sufficient* and *expectations as justified*, but in no case does the success of the strategy depend on the assumption of metaphysical physicalism. I therefore think that such a "de-ontologization" of the debate about the correct understanding of disposition expressions could only do good.

But back to the red rose. We have seen that the transition from disposition to manifestation correlates with a change in external circumstances. One deletes the light—the "Disposition red" is attributed to the rose; one switches it on again—the actual property "red" is attributed to the rose. This can be repeated at will. The manifestation of the quality "red", however, also depends on other external circumstances. In the extremely stimulating book of Kreißl and Krätz "Fire and Flame, Sound and Smoke", an interesting attempt, called "Rose Magic", is described:

If a rose is brought into an atmosphere with a high proportion of sulphur dioxide, the rose is bleached white. If you then wet the white rose with hydrochloric acid, it gets its red colour back. And if it is subsequently treated with ammonia, it changes colour from red to blue (cf. Kreißl and Krätz 1999, p. 220 ff.).

The colour of the rose depends on the way the cyanine dissolved in the cells of the rose petals reacts with these chemicals. That is, the actual property of the rose to contain cyanine opens the dispositions to appear colourless, red or blue depending on the environment. The fact that we consider the sentence "The rose is red" to be a completely trivial statement is obviously due only to the fact that our atmosphere is "normally" not soaked in sulphur dioxide, hydrochloric acid or ammonia.

The interplay of disposition and manifestation can, however, be taken a little further. At first it was said: Where no light, there no colour. But which colour we see sometimes depends on the light in which an object appears to us. Here, too, a small experiment can be carried out. First, you put a red rose in a green vase in front of a white background in white light. If you now illuminate this arrangement with green light, you will see that the background and the vase as well as the stems and leaves appear green, while the rose blossom appears in an indefinable colour, at least not red. Then you change the light and look at the whole thing in the red light. Now the flower remains red like the background, the originally green elements change colour from green to indefinable. Finally, you switch to blue light. The background appears blue, but everything else loses its original colour. If you were to experiment with rose magic, under different lighting scenarios, this would also result in completely different colour effects.

This suggests a new idea: perhaps colours are not at all genuine properties of things, but properties of light. This means that things have no colours, the light only makes them *appear coloured*. Whereby this again can mean different things:

Variant 1: Light itself is originally without colour; through contact with the object, it undergoes a change that produces an impression of colour.

Or

Variant 2: White light is composed of coloured rays of light, and things have the disposition to reflect colours of light.

Or

Variant 3: White light is made up of various components which have the disposition to produce colour impressions in humans.

3.2.4   Colour as an Actual Property of Light or as Its Disposition?

The first variant was negotiated in the history of science under the name "Modification Theory of Light", and René Descartes (cf. Descartes 1637) was its most famous representative. I will not go into this further, since the representatives of this theory are now extinct. Isaac Newton is associated with the other two variants [cf. Newton (1671/72) and (1704)]. Since some of his achievements in the field of optics continue to have an effect to this day, I would like to go into his relationship to colour in more detail [for the history of colour theory, see, e.g. Lampert (2000) and (2008)].

Contrary to Descartes, who at the beginning of his optical investigations presented a metaphysical, empirically unassailable model of the actual nature of light, Newton refused to answer the question of the true nature of light. Not because he had not thought of anything to do with it, but because he considered this question to be empirically inaccessible and thus "unphysical". In contrast to Descartes, Newton postulated that white light consists of many different components, the "rays of light". What these rays of light are is not said, but how they can be *represented*—namely by a physical quantity, the "angle of refraction", whereby the values of this quantity are to determine the difference of the rays of light in an unambiguous way. This means that the term "light beam" is indirectly defined by the *measurable behaviour of the quantity "refraction angle"*. In his first definition of "optics", Newton first states:

> I call the smallest part of the light that can be collected or emitted separately from the rest of the light, or that does or suffers something that the rest of the light does not do or suffers, ray of light. (Newton 1704, p. 5)

That sounds a little cryptic at first. What "does or suffers" then a ray of light ? The answer is essentially already given in the title of the optics given. Here it says, "Optics or treatise on reflections, refractions, diffractions and colours of light" "s". This means that the ray of light is reflected, refracted and diffracted, whereby the refractibility (Refangibility) plays a special role: Light rays can be distinguished from each other by being refracted in different ways in contact with a transparent medium. In this sense, for example, a "beam of light" has been technically well realized if, after passing through a prism, the light is not fanned out into a spectrum when it hits a screen, but lands at a point and the path of the light can be represented at least approximately as a line with an angle of refraction. In the second definition, the light beam is then determined as a physical object by a disposition:

> The refractibility of light rays is their disposition to be refracted or deflected at the transition from one transparent body or medium to another. (ibid.)

This dispositional way of speaking can be reconstructed quite well in the sense outlined above: sufficient experimental experience has been gained and this supports the expectation that light rays will show the same refractive behaviour under the same suitable conditions. Only the determining part, which is implicitly contained in the dispositional speech, causes some headaches. In our earlier examples of rose, glass, sugar and copper cables, the findings "this is a rose", "this is sugar", etc. could be confirmed by checking the current characteristics. But what about the "light"? "At Newton, "light" is defined as what behaves in a certain way under certain circumstances". The list of circumstances and behaviours is open and is determined solely within the framework of a theory, optics. A term characterized in this way is called a "theoretical" term, and the type of characterization is typical for physics and other sciences. At this point, Newton's contemporaries (and not only these) would have liked to have had information on the question of what light "really" was. But Newton already categorically states this in the first movement of his "Optik":

> It is not my intention in this book to explain the properties of light by hypotheses, but only to state them and to confirm them by calculation and experiment. (ibid.)

In other words, light and light rays are theoretical entities that initially only exist within the framework of physical optics: light is what physically behaves like light. How to define theoretical concepts in more detail is an important topic in the theory of science, which I cannot go into here in more detail.

For the measurement of the different refraction angles over which the different components of the light are fixed, it is of central importance that the spectrum, which results after the passage of a white light beam through a prism, results in a coloured image. There is a fixed correlation: light rays with the same refraction angle always show the same colour. This correlation facilitates the empirical measurement work enormously and at the same time suggests a new hypothesis, which Newton likes to ascribe to it: "White sunlight is a mixture of light rays of different colours".

In fact, Newton has posts in his first treatise on optics from the year 1672 which can be read in this sense. Here, it says:

> Colours are not acquired properties (qualifications) of light which, as is generally assumed, they have obtained by refraction or reflection on natural bodies, but peculiar and genuine properties which are different in different rays. (Newton 1671/72, p. 27)

So here, Newton turns against variant 1 and reveals himself as a representative of variant 2. But in the following decades, his methodological attitude sharpened, and in his optics from 1704, he states unmistakably in a "definition":

> The homogeneous light and the rays of light that appear red or rather which make objects appear red, I call reddening [in the original here stands the art expression 'rubrifick'] or red-generating; such rays of light that make the objects appear yellow, green, blue and violet, I call yellow-generating, green-generating, blue-generating, violet-generating, etc. And if I once speak of coloured or coloured rays of light, this is not to be understood in a scientific or strict sense, but in colloquial terms, according to the ideas that ordinary people would have if they saw all the experiments. Because strictly speaking, the rays are not colored. There is only a power or disposition in them to arouse the sensation of this or that colour. (Newton 1704, p. 80)

Newton is a physicist and physics deals with the world only to the extent that it can be represented by physical quantities. Such quantities are, e.g. length, time, mass, charge, angle of refraction or wavelength. In any case, "colour" is *not* a physical quantity and so physics does not have to remain colourless, but colour-free, i.e. the Newton of the later years is a representative of the third variant [similar to Campbell today (1969); overview of "dispositionalism" in Harvey (2000)].

In the sense of this variant, it is only a lucky circumstance that we have the ability to see colours; this helps enormously in the optics. But that does not make the colour part of the look. On the other hand, optics can do a lot to clarify the physical circumstances under which certain colour impressions occur in us. Goethe had always demanded that physics should take more care of this task, I will come back to it later. And, there is a wonderful book by Marcel Minnaert that shows how this task can be accomplished (cf. Minnaert 1992). Nevertheless, it remains the case that initially no more than a correlation between physical facts on the one hand and colour impressions in our head on the other can be determined.

3.2.5 Seeing Colours as a Human Disposition

This brings us finally to the question of what actually has to happen in our heads for us to have colour impressions. Even at school, we learn today that light is indeed composed of various components that are perceived as electromagnetic waves in a certain wavelength range. Light is the visible part of the wide spectrum of electromagnetic waves. The light is guided through the eye lens to the retina, where it encounters different photoreceptors. First, there are the so-called rod cells, which are more at the edge of the retina, are extremely sensitive to light and enable us to see in dark twilight or in the dark; rod cells, however, do not allow colour impressions. In addition to the rod cells, there are the so-called cone cells, which enable colour vision. They are located in the centre of the retina (visual pit, fovea centralis) and can be divided into three groups according to their particular sensitivity to light of a certain wavelength: The S type reacts particularly to light in the short-wave range (with a sensitivity maximum at 455 nm, "blue"), the M type is sensitive to light of medium length ("green", 535 nm) and the L type reacts mainly to long-wave light ("red", 570 nm). However, the receptors are not sharply adjusted to certain wavelengths, but have a large overlap range. . The light triggers chemical reactions at the receptors, but it is not the wavelength that is responsible for this, but the energy associated with it (which is contained in the photons). The chemical processes are again converted into electrical impulses, which then form the basis for further processing in the brain.

For a long time, this picture was the basis of a reductionist understanding of colour vision: From the (almost) continuous light spectrum, three colour-specific wave bands are filtered and from these, as in the three-colour technique of television, the variety of colour impressions is synthesized. In this picture, "colour" is regarded as the sensation triggered by electromagnetic oscillations in the visible range.

But things are not that simple. Because as soon as the receptors have done their transformation work, the deeper regions of the retina and the brain begin to process these signals according to their own rules. Nerve cells that derive the signals from the retina regroup them into the four components red, green, blueand yellow, whereby the components red, green, blue and yellow do not allow any colour transitions. This means that a four-colour theory of colour production applies to the brain, although there is no receptor at all for the colour "yellow" at the cone cells.

A second important and own brain achievement is the maintenance of the colour constancy. If the simple reductionist image were correct, we would actually have to see our surroundings in a constantly changing colour scheme, since the relative wavelengths of sunlight change over the course of the day. That is not going to happen. Although our photoreceptors are addressed to varying degrees, the brain insists on always getting the same colour impression.

This human ability has greatly preoccupied the American physicist and entrepreneur Edwin Land and has led to a series of amazing experiments. All these experiments produce one result: a surface retains its colour impression even when illuminated with light of different spectral composition. Only condition: the coloured surface is embedded in a coloured environment. This can only be understood in such a way that the brain apparently not only records the distribution of the wavelengths of light, but also the light reflexivity of the surface (because this is retained at different exposures). This is then placed in relation to the reflectivity of the surrounding surfaces. If the relational ratio remains the same over time, so does the colour impression, even if the lighting conditions have changed dramatically (Solomon and Lennie (2007) give a brief overview of the physiology of vision).

The great physicist James Clark Maxwell, who more than 140 years ago, regarded colour as a purely subjective perception and therefore demanded: "The science of colour must be regarded as a pure topic of brain research", is right. "The science of colour must therefore be regarded as essentially a mental science." (Maxwell 1871, p. 13). No, this assertion also goes too far. On the contrary, previous observations suggest the conclusion that the phenomenon of "colour" cannot be justified by a simple "either objectivistic or subjectivistic" approach. The colour of an object is neither a property of the object nor a property of light nor a simple sensory impression. The "color" is a kind of index in which our brain *summarizes* a whole series of physical and physiological data *into a visual quantity*. We humans have this ability to index the environment in common with many animals, even though this ability is very different (fish can adapt their colour sense to different water depths, bees can switch off colour vision when flying fast to relieve their brain of the otherwise necessary data processing). It looks as if man, with the help of colour impressions, is structuring his space of perception and experience in the pre-linguistic realm in a similar way as he does in the linguistic realm with the help of predicators. In both cases, the discriminatory powers are stimulated, but not imposed, by the outside world. And it is above all the linguistic abilities of the human being that allow him to *reflect on* the achievements of his visual discrimination. and possibly *relativize* it.

3.2.6 From Philosophy to the Phenomenology of Colour

So people do not simply perceive colours, they actively ascribe symbolic meaning to them, associate them with associations that indirectly produce psychological effects, produce colours technically, use them artistically and evaluate their handling according to aesthetic aspects. This is discussed in more detail in Sect. 3.3. Although Wittgenstein has addressed these real-life questions in his typical cursory manner (Wittgenstein 1984), contemporary analytical philosophy has not spun his thread further. One insists that, in epistemological terms, a colour impression is merely the result of passive perception and denies the question of the independence of this perception through a commitment to a metaphysical physicalism. Fatal is only that philosophers can neither redeem this confession themselves practically nor give examples for a successful reduction of mental phenomena in the broadest sense to physical events. So it remains with a defiant and argumentation-free "Here I stand and will not change".

In its programmatic approach, Goethe's theory of colour, which is otherwise little appreciated today, is much closer to a practical, technical and artistic approach than to contemporary philosophy, and I would like to conclude by briefly discussing it. In his theory of colours, Goethe has programmatically expressed that the phenomenon of "colour" can only be successfully approached in its manifold physical, psychological, cultural and aesthetic aspects if this is done in an interdisciplinary network. Goethe's theory of colour is an attempt to systematically capture the colour phenomena that we encounter in everyday life in different contexts. The explanation of these colour phenomena is then assigned to the sciences (physiology, chemistry, physics, psychology, cultural sciences) which are at home in these contexts (cf. Goethe 1808).

So it is consistent when Goethe rejected a strict dichotomy in "either color is objective and thus physics or it is subjective and thus physiology" as infertile. He thus stood in contrast to many contemporaries who were used to thinking in dichotomies and who therefore met his reflections with incomprehension.

In order to be accessible to scientific treatment, however, the phenomena must be removed from their everyday environment and transferred "into the chamber" (today one would say "into the laboratory") where they can be studied under controlled conditions. This transfer occurs through the "attempt", as Goethe described this in his essay "Der Versuch als Vermittler von Objekt und Subjekt" (cf. Goethe 1823). The "experiment" is—in modern terms—a real simulation in which effects can be created and studied by varying various parameters, but it is not an experiment. Goethe was expressly prepared to assign the explanatory claims of the phenomena to the specialist sciences, as his achievement merely claimed to have compiled the phenomena himself in a systematic manner [a phenomenological analysis of colour phenomena in Goethe's sense can be found today, for example, in Nussbaumer (2008)]. All this is clearly expressed in the introduction to the didactic part of colour theory. Only with regard to Newton's optics did he, as is well known, have his reservations, since he wrongly assumed that Newton had ontological premises which made this theory unsuitable for him to explain the phenomena (this vehement, but in substance inaccurate criticism has prevented many people to this day from dealing more closely with the programme of the theory of colours, although the argument with Newton is actually irrelevant for the theory of colours as a whole; see Kötter (1989) for details).

For the practical and technical use of colour, Goethe considers philosophical rigorism to be of little help. It is not the question of *what* colour is that is decisive, but *how* it appears. "For here, too, we have no choice but to repeat that colour is the lawful nature in relation to the sense of the eye. Here, too, we must assume that someone has the meaning, that someone knows the influence of nature on this meaning; for with the blind one cannot speak of color" (Goethe 1808, p. 324). Actually, today everyone who is practically involved with colour, be it in technology, design or art, follows Goethe's interdisciplinary programme, even if they are generally not aware of these historical roots. Section 3.3 shows this aptly.

Bottom Line

We have seen that attempts to assign colour a clear ontological status as either a physical or mental entity have not led to convincing results. "Colour" is a phenomenon that can best be approached by different ways of life and science, and each of these approaches has its own philosophical problems, which require specific epistemological, linguistic-philosophical or scientific-theoretical reflection in order to be solved. The philosophical "general problem solver " will certainly not exist for the problem complex "color".

3.3 Colour Concepts for Workstations

What contribution does colour make to the furnishing of humane, functional work-places? In the following, the meaning of colour for people and space is defined, followed by a procedure for colour design that leads to physiologically relieving, psychologically beneficial colour concepts.

3.3.1 Does Pink Red Cool Hot Minds?

A few years ago, prisons in the German-speaking world opened cells for emotionally heated prisoners with "Cool Down Pink" painted. They referred to an investigation which allegedly proved that this rose red had a calming effect. Will the office of the future be pink? Probably not—prisoners have found the painting humiliating, and the positive experiences quoted in the media have proved to be anecdotal narratives. Most of the pink cells were painted white again, and other branches were spared the pink detour.

The "neutral" white office has been booming for years. Is that the best solution? The white coat of paint is a safe haven for many and the lack of substantiated state-ments about colour effects that can withstand critical examination cannot invalidate the tendency towards white. The often cited equation of white with clarity and purity, but also cost reasons, seems to speak for the white office world. Nevertheless, there are objective arguments against white and for more contrasting working environments.

3.3.2 What Do We Need Paint for?

What brings colour to people and spaces (Figs. 3.1 and 3.2)? Le Corbusier quoted Fernand Léger and wrote: "Man needs colour to live; it is as necessary an element as water and fire" (Le Corbusier, quoted after Rüegg 2015). What was he thinking? Without water, we die of thirst. Without fire, we will freeze to death. No paint?

Our sense of sight has specialized in perceiving a range of solar energy known as "light" as colour on surfaces. This incredibly refined interplay of light, surfaces and perception is what "colour" is. What happens when we see colour is a miracle of evolution:

1. Light hits an object, a rose blossom serves as an example.
2. The pigments and the surface structure of the object change the light both in its orientation and in its composition.
3. The photoreceptors in the eye absorb the altered light.
4. Nerves transmit the information.
5. A cascade of chemical signals is triggered in different regions of the brain.
6. We see in the garden in some distance a landscape with red rose.

Fig. 3.1 *Meeting room with soft grey to* grey contrasts; Charles-Édouard Jeanneret-Gris (Le Corbusier), 1912, colours correspond to original colours

We perceive the red rose, together with the green grass, the sky and numerous round shining pearls. We perceive the pearls as dew drops in the morning and raindrops in the afternoon. Later we enter the garden again. The rose is now almost brown and the meadow dark grey, but we evaluate the picture as a red rose on green grass at dusk. This incredibly precise object recognition, which ranges from individual roses and blades of grass to complex landscapes or building sequences, is possible because we see light on surfaces as colour and can use shifts in the light and colour spectrum as a source of information. Colour vision provides us with information at intervals of about 13 ms (cf. Potter et al. 2014) that makes objects, backgrounds, weather conditions, times of day and much more visible to us.

No paint? Without colour, we lose our visual orientation. in space and time (Figs. 3.3 and 3.4) (Trautwein 2014, p. 105).

Being able to see colourful and non-colourful colour differences enables us to visually identify objects, dangers and situations (cf. Gegenfurtner and Kiper 2003).

Fig. 3.2 An office with RAL 9010 on the walls and hard light-dark contrasts

Fig. 3.3 A picture with only uncoloured light-dark contrasts

Fig. 3.4 The same picture with colourful and uncoloured chiaroscuro contrasts

3.3.3   Seeing Means Seeing Colour Differences

The list of information we perceive as colour effects on surfaces is impressive: places, landscapes, forms, colours, spaces, foregrounds and backgrounds, objects and contexts, shadows, movements and speeds, depths and distances, materials, atmospheres, times of day and seasons (cf. Chirimuuta 2015). The snow diagram in Fig. 3.5 can serve as an illustration. The landscape can easily be recognized as the shadow of a tree on the new snow cover. In the pre-consciousness a comparison of information between optical stimuli and images of experience has taken place and we also know that it is winter and the sky is bright blue. The lengths of the shadows tell us that skiing will soon be a thing of the past. Feelings of pleasure and displeasure connect with this (and any other consciously or unconsciously perceived) image. It is not the blue tone or any other colour that is decisive for the emotions, but the overall composition and how we instinctively interpret it.

   To clarify the statement that it is not the recognition of individual colours that is important, but the emotional evaluation of the overall situation: we all immediately recognize a strawberry and whether it is greenish-red and immature, deep red and ripe or purple-red and overripe. Purple-red does not generally appear, but only in this context unappetizing. As a colour concept maker, the lever for this instinctive reaction can be applied to the overall visual context. The working spaces of the future will not be white, pink or blue, they will form pleasant landscapes.

Fig. 3.5 Snow shadows—everyone recognizes the situation immediately, although everyone sees the blue differently

3.3.4 The Design of Soothing Colour Landscapes

How do you create a working environment that appeals to immediate feelings of pleasure rather than displeasure? When we enter a room, a firework of sensory impressions arrives in our subconscious. A complicated variety of information is sifted, and a biological reaction is triggered that takes place long before we step back and think about what we should think of space. Staying or fleeing, laughing or crying, acting or persevering, feeling pleasure or displeasure—we do not think about this, it happens unconsciously with us (cf. Robinson and Pallasmaa 2015, p. 20; Gladwell 2005; Gordon 2015, p. 11). What does a workspace look like that appeals to positive emotional experiences and works against physiological fatigue or boredom? The research results of recent years can be interpreted as evidence that the fulfilment of the following conditions leads to spaces that invite people to stay (Fig. 3.6):

1. The room offers orientation. This means that important objects in the room can be easily recognized, and unimportant ones are designed unobtrusively.
2. The colour contrasts, and shadows in the room are soft. The colours change subtly with the light during the course of the day. These material properties create emotional connections to the harmonious colours of nature.
3. The room is in harmony with visual common sense. The soil colour can be earthy or stony, but not sky blue, sea blue or bright pink. Red seems economical at all, yellow is neither to be used on the ground nor on shaded surfaces, not everything must be white, etc. The systematic discussion of these rules goes beyond the scope of this article. For further information, please refer to specialist seminars devoted to these topics.

Fig. 3.6 The sky-blue wall seems to recede; what does the wall good would irritate the floor; sky-blue floor colours offend against visual common sense

3.3.5 The White Office as a Problem Case

What happens physiologically when entering a radiantly white room? Imagine a brightly lit room with white walls, ceilings, window frames and curtains (Fig. 3.2). You enter the room and see the white walls first. White things want to be seen. Our visual sense perceives both bright colours and clear colour contrasts most quickly. This is probably why Weiß seduces so many architects—they suspect that white forms appear particularly clear and distinct. White environments, however, in fact have two decisive disadvantages. The first is that our slits contract due to the dominant brightness. The light supply is reduced so that the photoreceptor cells are not saturated. As a result, darker objects, such as furnishings, books, monitors and office neighbours, appear unclear and darker (Fig. 3.7).

The second disadvantage is related to the lack of variety. Every monochrome environment makes orientation more difficult for us, whereas the monochrome white environment makes a double effort on our perception due to its brightness. Brightness and contrasts attract our attention, but uniformity balances out differences. It becomes more difficult to recognize what is important and what is unimportant in the visual field. In the example (Fig. 3.2), the walls are designed to catch the eye of the room!

Fig. 3.7 Two identical yellow fields, the yellow field in the white environment is difficult to see, clearly visible in the light grey environment

In the very white room, every white surface demands our immediate attention. It is as exhausting as monotonous music that is too loud.

Moderate white or even light grey environments are more physiologically friendly than clinically white ones. The viewing slits remain open, the objects in the room appear clear and colourful, and the contents of the room do not have to struggle with the walls for attention. At the same time, however, the environment must not be monochrome light grey, which would also be too undifferentiated. It depends on the design of the contrasts (Fig. 3.8).

3.3.6 Three Steps to the Perfect Colour Concept

Using colour in such a way that it creates a pleasant atmosphere and clarifies the formal language of the architecture is a three-stage process. First, a light base colour and a matching dark shadow colour are defined to create the desired atmosphere (Fig. 3.9). In the second step, light and dark surfaces are arranged in the room in order to increase its depth effect, to determine the direction of movement of the room user and to differentiate surfaces and forms in the hierarchy of their meaning (Fig. 3.10). Finally, contrasts are reinforced by dynamic, therefore rich, colourful colours and spaces are individualized, with a red wall for example or a blue niche (Fig. 3.11). As this is of utmost importance for the atmospheric quality of the room, the following will concentrate on the choice of the light basic colour and the matching dark colour.

White surfaces can be bright or dull, inviting or rejecting. Two properties determine the effect of colours in connection with light and form: on the one hand, their remission spectrum—the area in which light wavelengths are absorbed or rejected by colour pigments—and on the other, the way in which non-absorbed or remitted light

Fig. 3.8 The background colour used over a large area may be light, but it should not overshadow the room contents; here the colour KT 200.143 Ombre naturelle pâle provides unobtrusive backgrounds; the room contents become an eye-catcher, not the walls

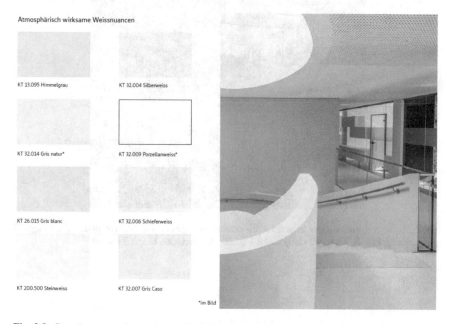

Fig. 3.9 Step 1—atmosphere: choose the light colour, it shapes the atmosphere in the room; KT 32.014 Gris nature and KT 32.009 Porzellanweiss

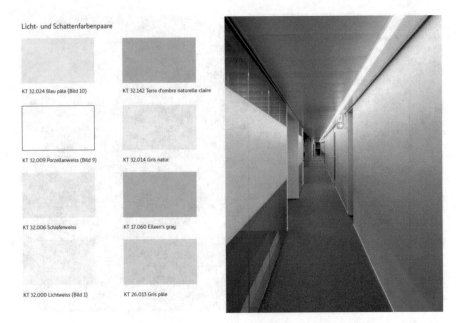

Licht- und Schattenfarbenpaare

KT 32.024 Bleu pâle (Bild 10)

KT 32.142 Terre d'ombre naturelle claire

KT 32.009 Porzellanweiss (Bild 9)

KT 32.014 Gris natur

KT 32.006 Schieferweiss

KT 17.060 Eileen's gray

KT 32.000 Lichtweiss (Bild 1)

KT 26.013 Gris pâle

Fig. 3.10 Step 2—differentiate: choose the colour to the light main colour for areas that are to be hidden from attention

Fig. 3.11 Step 3—set colour accents: individualize the spaces and prevent budding feelings of arbitrariness; KT 43.7 Vert vif (left) and KT 03.001 Ultramarine blue Y3 (right)

is rejected. Remitted light is either reflected or scattered by surfaces. Since white colours absorb little light, the type of remission, reflection or scattering is decisive for the effect of white architecture. Light-reflecting surfaces sparkle and shine, while light-scattering surfaces seem flat and intangible. The difference between reflection and scattering corresponds to that between the glitter of fresh powder snow and the flat effect of compacted snow layers. The reflection of the new snow from crystalline particles produces lively shadows, the scattering caused by crushed particles diffuse grey shadows. The white pigment most frequently used today, titanium white, is micronized to a defined particle size at the end of its manufacturing process. The grinding process to produce particles, ideally between 0.2 and 0.3 μm in size, produces small, homogeneous pigment particles that increase the opacity of a colour but eliminate sparkling mirror effects. Titanium white and its relatives such as nickel titanium yellow and other mixed-phase oxide pigments therefore scatter light from their surfaces and have an intrusive effect. In contrast, natural pigments such as chalk, lime, marble, kaolin and umbra tones show mirror effects due to their crystalline structure. A white made of pure champagne chalk reflects and modifies light in a completely different way than titanium white. Light penetrates into the colour layers of the natural limestone and is reflected back from different depths. The surface has the tactile softness and luminosity typical of natural pigment colours. In the architecturally important white and pastel areas, depth is gained by dispensing with titanium dioxide, and in other shade areas by using colours made mainly of natural pigments. They enhance the value of architecture and improve its environmental balance. In other words, the typical light in a room only comes from the effect of light on all surfaces, and a natural room light only comes from natural pigments.

After a light colour has been selected, a natural "shadow color" is selected that matches it. Which surfaces should carry the darker tone? Three questions help you make your decision: Where should the impression of room depth be reinforced? Along which axes do the movements run through space? What is unattractive and should be hidden from view? A narrow room continues to have an effect if the shaded wall is painted darker and the wall standing in the light brighter. Through this differentiation, the bright wall is clearly visible, while the dark wall attracts little attention and even seems to retreat. The stronger the contrast, the more dramatic the effect.

Movement through complicated room situations can be controlled by the effect that bright surfaces are perceived faster than dark ones. The rapid physiological reaction to brightness contrasts enables the designer to emphasize surfaces or room parts differently. Walking through architectural ensembles, we are attracted by window openings, illuminated surfaces and bright rooms. Dark surfaces and rooms, on the other hand, are only perceived at second glance. With the help of this surface differentiation, low ceilings, excessively narrow passageways or other unsightly components can be distracted. By keeping the walls of a staircase in shadows and allowing conference rooms on the upper floor to shine brightly, guests are led up the stairs into the light. If they succeed, they will feel just as naturally attracted by the illuminated conference rooms as walkers in a winter night by the warm glow of the windows.

How dark may the darkness be? Usually, rooms in which people spend the day are painted brightly overall. Differences in brightness of at least 15% ensure a differentiation that makes colour concepts understandable. In this context, it should again be pointed out that people and works of art appear fresher in front of walls in muted tones.

How do you design spaces that invite you to work? That was the question at the beginning of the section. "Beauty speaks like an oracle", said Luis Barragán in a speech he gave in 1980 at the reception of the Pritzker Prize (Barragán 1980). In this, he described the concepts of silence, seclusion, serenity, joy and closeness to nature as the guiding principles of his poetic architecture. Pigmented colours, which sensitively support the interactions between light and volume, offer a suitable method for the design of architectural ensembles in the sense of Barragán, because light, dark and colourful colours used in a differentiated way create spaces in which creativity can unfold freely.

Literature

Armstrong, D. M. (1969). Colour-realism and the argument from microscope. In R. Brown & C. D. Rollins (Eds.), *Contemporary philosophy in Australia* (pp. 119–131). London.

Barragán, L. (1980). The Pritzker architecture prize 1980 laureate acceptance speech. http://www.pritzkerprize.com/sites/default/files/file_fields/field_files_inline/1980_Acceptance_Speech.pdf. Access March 28, 2014.

Byrne, A., & Hilbert, D. R. (1997). Colors and reflectances. In A. Byrne, & D. R. Hilbert (Eds.), *Readings on colors, vol. 1: The philosophy of color* (pp. 263–288). Massachusetts: Cambridge.

Campbell, K. (1969). Colours. In R. Brown & C. D. Rollins (Eds.), *Contemporary philosophy in Australia* (pp. 132–157). London.

Chirimuuta, M. (2015). *Outside color: Perpetual science and the puzzle of color in philosophy.* Cambridge, Mass: MIT Press.

Cuykendall, S. B., & Hoffman, D. D. (2008). *From color to emotion. ideas and explorations.* Irvine, CA: University of Irvine Press.

Damböck, C. (2013). *German empirism—Studies on philosophy in German-speaking countries 1830–1930.* Springer.

Descartes, R. (1637). Discours de la Method pour bien conduire sa raison, & chercher la verité dans les sciences. Leiden (*German: Entwurf der Methode. With the dioptric, the meteors and the geometry.* Edited by Chr. Wohlers. Hamburg 2015).

Dorsch, F. (2009). *The nature of colours.* Frankfurt/M.

Gegenfurtner, K. R., & Kiper, D. C. (2003). Color vision. *Annual Review of Neuroscience, 26,* 181–206.

Gladwell, M. (2005). *Blink. The power of thinking without thinking.* London: Penguin Books.

Goethe, J. W. (1808). On the theory of colours. Didactic part. In: Goethe's Works HA, vol. 13, Munich 1981, pp. 314–523.

Goethe, J. W. (1823). The experiment as mediator of object and subject. In Goethe's Works HA, vol. 13, Munich 1981, pp. 10–20.

Gordon, G. (2015). *Interior lighting for designers* (5th ed.). Hoboken, NJ: Wiley.

Hardin, C. L. (1986). *Color for philosophers.* Indianapolis: Unweaving the Rainbow.

Hardin, C. L. (2014). Color qualities and the physical world. In: E. L. Wright (Ed.), *The case for Qualia* (pp. 143–154). Massachusetts: Cambridge.

Harvey, J. (2000). Colour-dispositionalism and Its recent critics. *Philosophy and Phenomenological Research, 61,* 137–153.

Jackson, F., & Pargetter, R. (1997). An objectivist's guide to subjectivism. In: A. Byrne & D. R. Hilbert (Eds.), *Readings on colors, vol. 1: The philosophy of color* (pp. 67–80). Massachusetts: Cambridge.

Kötter, R. (1989). Newton and Goethe on the theory of colour. *German journal for philosophy, 46,* 585–600.

Kreißl, F. R., & Krätz, O. (1999). *Fire and flame, sound and smoke.* Weinheim.

Lamb, T. D. (2016) Why rods and cones? *Eye, 30,* 179–185.

Lampert, T. (2000). *On the science theory of color theory. Tasks, texts, solutions.* Bern.

Lampert, T. (2008). Newton vs. Goethe: Farben aus Sicht der Wissenschaftstheorie und Wissenschaftsgeschichte. In: H. Bieri & S. M. Zwahlen (Eds.), *"Drink, o eyes, what the eyelash holds, …". Colour and colours in science and art* (pp. 259–284). Bern.

Maxwell, J. C. (1871, May 4). On colour. *Nature,* 13.

Minnaert, M. (1992). *Light and colour in nature.* Basel.

Newton, I. (1671/72). A new theory about light and colours. In: Phil. Trans. No. 80, pp. 3075–3087 (English: *Newton's theory of prismatic colours* (Ed. J. A. Lohne, B. Sticker), Munich 1969).

Newton, I. (1704). Opticks: Or, a treatise of the reflexions, refraction, inflexions and colours of light, London 1704, 21717, 31721, 41730 (*English: Optics or Treatise on Reflections, Refractions, Diffraction and Colours of Light.* (Ed. W. Abendroth), Leipzig 1898, Frankfurt/M 1996).

Nüchterlein, P., & Richter, P. G. (2008). Space and colour. In P. G. Richter (Ed.), *Architectural psychology* (pp. 209–231). Lengerich: Pabst.

Nussbaumer, I. (2008). *The theory of colour. Discovery of the messy spectra.* Vienna.

Potter, M. C., Wyble, B., Hagmann, C. E., & McCourt, E. S. (2014). Detecting meaning in rapid pictures. *Attention, Perception, & Psychophysics, 76*(2), 270–279.

Robinson, S., & Pallasmaa, J. (Eds.). (2015). *Mind in architecture. neuroscience, embodiment, and the future of design.* Cambridge, MA: The MIT Press.

Rüegg, A. (2015). *Polychromy architecturale. Le Corbusier's color keyboards from 1931 and 1959.* Basel: Birkhäuser.

Schleichert, H. (1975). *Logical empirism—The Viennese circle.* Paderborn: Wilhelm Fink.

Solomon, S. G., & Lennie, P. (2007). The machinery of colour vision. *Nature Review Neuroscience, 8,* 276–286.

Stone, N. J., & English, A. J. (1998). Task Type, poster, and workspace color on mood, satisfaction, and performance. *Journal of Environmental Psychology, 18*(2), 175–185.

Thompson, E. (1995). *Colour vision. A study in cognitive science and the philosophy of perception.* London, New York.

Trautwein, K. (2014): Black-Black. Zurich: Lars Müller Verlag, Uster: kt.COLOR AG.

Trautwein, K. (2017). Colour concepts with light and shadow colours. Seminar contents and dates at www.ktcolor.ch. Uster, Switzerland.

Vetter, B., & Schmid, St. (Eds.). (2014). *Dispositions.* Frankfurt/M: Texts from the contemporary debate.

Wittgenstein, L. (1984). Remarks about the color. In: *Works edition* (vol. 8, pp. 7–112). Frankfurt/M.

Chapter 4
Adequate Office Interior Design

Birgit Fuchs, Thomas Kuk and David Wiechmann

4.1 Introduction

4.1.1 The Office and Its Significance

Setting up an office is the last step in a long process chain. Nevertheless, it is the step closest to the human factor that is decisive for the well-being and motivation of employees. The equipment reflects the soul of a company, which is an important part of the corporate identity and thus the external impact. The spatial arrangement in companies reflects the organizational hierarchy and the department affiliation.

The basic aim of planning is to combine function and emotion in a target-oriented way. This means that a workplace must take into account the requirements of a particular task and at the same time meet the individual needs of its user.

4.1.2 The Topics in the Contemporary Office

In the course of time, the office has undergone many developments, which have been characterized by new findings in ergonomics and organizational development. The

B. Fuchs (✉)
Steelcase Inc, Munich, Germany
e-mail: bfuchs@steelcase.com

T. Kuk
Spielmann Officehouse GmbH, Kronberg, Germany
e-mail: thomas.kuk@designfunktion.com

D. Wiechmann
Kinnarps GmbH, Worms, Germany
e-mail: david.wiechmann@kinnarps.de

© Springer Nature Switzerland AG 2020
W. Seiferlein and C. Kohlert (eds.), *The Networked Health-Relevant Factors for Office Buildings*, https://doi.org/10.1007/978-3-030-22022-8_4

aim has always been to ensure maximum efficiency. Both in terms of productivity and cooperation but of course also with regard to cost structures.

4.1.3 The Many Facets of Open Space

The classic single room is now considered a privilege from days gone by and has largely served its purpose as a model. Nevertheless, astonishingly many people still want their own office—in contrast to the popular notion of the cool marketing consultant, who is using a laptop in the creative zone, the agency or the beach.

Half the generation Y and more than 65% of the generation X (born between 1964 and 1980) would like to have a permanent place in the office. They want their own desk and would prefer to be involved in the design of their office. Only around five per cent of all respondents can imagine asking in the morning whether there is a desk available for them in the office and where. This was the result of a study by Savills and CCL (Savills, CCL/Consulting Cum Laude 2016).

When it comes to open space concepts, it should be considered that work has become much more complex and multi-layered. This concept also has to meet different requirements. The key is a mix of possibilities tailored to needs from areas for quiet, concentrated work to communication-promoting islands for creativity. Rest areas and think tanks must also be included in the planning.

Flowing boundaries are created with spaces that seamlessly blend between work and relaxation—a pleasant environment should be created in which employees feel invited to think outside the box in a relaxed manner.

4.1.4 Participation in Workplace Design

In contrast to the past, today more value is placed on individual needs and the well-being of employees. Acoustics, light, textiles or even colours are just some of the factors. Employee surveys provide insights into work processes, preferences and habits. They are transferred to requirement profiles for new work environments. Mock-ups enable the workforce to test a room concept and help decide whether they want to work in such an environment.

At the same time, a picture emerges with the need for individual workplaces. Many companies no longer have a workplace for every employee, and everyone has to find a free place to work.

4.1.5   Many Generations in One Company

From the baby boomers (born in the 1960s/1970s) through generations X and Y to generation Z, born around 1995, today four generations often work simultaneously in one company. This also plays an important role for working environments and is reflected in the "Green Paper Working 4.0—Thinking Work Ahead" of the German Federal Ministry of Labour and Social Affairs. It says, among other things: "The central objectives for age- and age-appropriate work are to create good and motivating working conditions […] and to protect and promote the health of employees" (German Federal Ministry of Labour and Social Affairs 2016, p. 26).

This means that the baby boomer, who grew up with a typewriter and folders, has to find himself just as much as the digital native, who shares his files in the cloud and communicates virtually.

4.1.6   War for Talents

The changing demographic structure of the population and the development towards a networking, highly mobile society will shape the German labour market in the coming years. Knowledge workers will be more sought after and at the same time will be more demanding than in the past. For a long time, companies were only interested in the capital, now the factor labour is also becoming an important resource. The so-called war for talents ensures that when choosing a job, not only the job itself must be right but also the circumstances. And that also includes an attractive workplace.

4.1.7   The Future of the Office

The mentioned report by Savills and CCL from 2016 has shown that employees have no concrete ideas of the office of the future. It is similar to the famous quote from car manufacturer Henry Ford: "If I had asked people what they wanted, they would have said faster horses". It can also be assumed in the future that industries and working methods will influence the office layout and that there are many approaches for different room concepts.

In the end, it will be a little bit of everything. Working in the company promotes loyalty to the employer and contact with colleagues, but does not exclude the home office if necessary. Room concepts include flexibly to the respective need. So, the workplace area becomes a creative zone without design borders: The use of stylish worlds to create garage spirit (Rococo, Biedermeier, stucco or even puristic; comfortable in special areas), individual work in a smooth transition to work in teams, the canteen area transforms into additional workstations if required. This also places special demands on office furnishings . They must be mobile, reusable and multi-functional in order to meet future requirements.

Working environments of the future no longer only offer pure function, but also experience, for example, in the form of different materials or visual impressions. The mega trends such as health and well-being will also manifest themselves in the workplace and go beyond ergonomic sitting or height-adjustable tables. Indoor climate, lighting and acoustics are becoming increasingly important (light, air, noise and body).

4.2 Ergonomic Office Furniture for the Workplace

4.2.1 Workspace or Space for Work?

The dynamic environment of an office encourages the employee to work in different postures and to move within the office. Movement is integrated into the office auto-matically. Ergonomics forms the methodical basis in all areas of setting up working environments. It refers to classical factors such as posture, movement, indoor cli-mate, light and acoustics, but also takes the subjective well-being of the employees seriously in terms of aesthetics and atmosphere.

The rapid change in the world of work has resulted in a huge paradigm shift in terms of work location, length of stay and type of task. It is certainly good if the ergonomic evaluation of workplaces and their implementation also include these factors.

The design of the personal workplace and the furniture equipment are defined very strongly by the activity of the individual. A distinction is made between the workstation which mostly consists of PC, table and chair, so-called *territorial working* from the workplace, which is available to all employees, to *non-territorial work*. This also results in a completely new aspect from the point of view of risk assess-ment ("The assessment of working conditions must identify all hazards and stresses which can have a negative impact on the health of employees"). All stresses—phys-ical, visual and psychological—must be taken into account (Deutsche Gesetzliche Unfallversicherung e. V. 2015, p. 17), in which conference and project rooms as well as lounges are used like a workplace and the actual computer workstation takes up less space, in the literal sense of the word. The choice of the territorial or non-territorial workplace is consequently a matter for the individual. In principle, a distinction is made here between *residents* and the *mobile workers*.

Rules for ergonomic office design are prescribed by law by the European standards and are checked in Germany by the professional associations in their application. The statutory minimum requirements and the additional ergonomic recommendations, for example, are presented in an understandable and comprehensible way for every user in the information provided by the German statutory accident insurance 215-

410 of the Administrative Employer's Liability Insurance Association (Deutsche Gesetzliche Unfallversicherung e. V. (DGUV) 2015).

In this chapter, it will be worked out how ergonomic office furniture can be used for and space design can make an important contribution to the physical well-being of employees in the office ("relationship prevention is about health prevention with regard to workplace design, the workplace, work equipment and the other working environment […]"): https://www.arbeitssicherheit.de/service/lexikon/artikel/verhaeltnispraevention.html), but also how employees can make a contribution to maintaining health through their behaviour in the working environment and the proper use of work equipment ("Behavioural prevention concerns prevention with regard to the behaviour of the individual at work and in connection with work". The aim of behavioural prevention is the "avoidance and minimisation of certain health-risk behaviours and psychological complaints […]", which is based on the individual person himself: https://www.arbeitssicherheit.de/service/lexikon/artikel/verhaltenspraevention.html). The focus is not so much on the dogmatic implementation of law and order, but rather on the observance of health-promoting measures to motivate and keep people healthy.

The well-being of employees can be achieved if factors such as colour concepts for furniture and other objects such as carpets, walls, etc. are also derived from a concept.

4.2.2 The Territorial Workplace—Minimal Movement in the Tightest of Spaces

Employees who have to work at their workstation need a useful strategy on how they can provide variety for their bodies and concentration. This is achieved by frequently changing sitting, standing and moving (Fig. 4.1).

Additional ways to the printer, to the waste-paper basket, to the coffee machine and the conversation with colleagues can help the body and soul on the jumps again. And sometimes breaks are more efficient than you think. Sensibly designed office

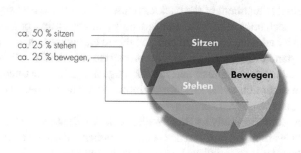

ca. 50 % sitzen
ca. 25 % stehen
ca. 25 % bewegen,

Sitzen
Stehen
Bewegen

Fig. 4.1 Sitting, standing and moving

furniture can do its job while working on the PC. These are, for example, the motion-promoting chair and a sit-stand workstation with a height-adjustable table or a table with an additional standing desk.

Those who do not move are in a vicious circle because long monotonous sitting leads to reduced blood circulation, which supplies the muscles with less oxygen. After some time, this leads to muscle hardening, tension and pain. If you are in pain, you tend to be more gentle, and the one-sided strain of which can ultimately lead to muscle hardening again and ultimately to muscle atrophy. The musculature loads thereby the spine and beyond that intervertebral discs, tendons and ligaments.

The reduced blood circulation also results in fatigue and reduced ability to concentrate, so that work is slower and the frequency of errors increases (cf. Tsunetsugu et al. (2007): visual stimulation of rooms with different quantities of wood—in 45% of the rooms a clear decrease in diastolic blood pressure and a clear increase in pulse were observed; this room tends to have the highest score in subjective comfort). So, what should be taken into account when setting up the territorial workplace?

4.2.3 A Good Office Chair

The task of the office chair is to optimally support the body and to guide the employee in his or her natural movements at work. The chair must ensure that no incorrect posture is adopted, and it should avoid prolonged monotonous sitting.

The three main disciplines are decisive criteria for the development of an office chair: Anatomy (the human body), anthropometry (the dimensions of the body) and biomechanics (the natural sequence of movements).

The five most important criteria for a good office chair

Function/Ergonomics
The office chair should have a harmoniously coordinated synchronous mechanism. This means that the seat and backrest move forwards and backwards in the same proportion as the body. The employee is thus in contact with the backrest in every sitting position (Fig. 4.2).

The individual adjustment of the backrest tension ensures the best support for the back. A good mechanics allows the relaxed sitting in any sitting position for every person, optimally adapted to his size and weight (in ergonomics, size and weight of the person are standardized by the EN). This is referred to as the 5th to 95th percentile, i.e. the 5% smallest and lightest, and the 5% largest and heaviest people living in Europe are the exception to the rule; for optimum adaptation, (special equipment or production should be chosen here).

At the latest since the study by Prof. Wilke et al. (1999) at the University of Ulm, sitting in a relaxed reclined position has been recommended because this reduces the pressure on the intervertebral discs and relieves the muscles.

Fig. 4.2 Office chair: synchronous mechanism

- The back is supported in its natural posture, and yet the movement of the upper body supplies oxygen to the muscles and nourishes the discs in the lumbar zone. Due to the good blood circulation, the concentration remains longer.

Backrest

The shaping in the lumbar area must ensure large-area support of the lower back without pressure points. Individual positioning can be achieved by adjusting the height of the lumbar support (Fig. 4.3).

- The spinal column remains supported in its natural position, for example, a hunchback is avoided. This means that the intervertebral discs are evenly loaded and retain their elasticity and function for longer.

Seat

An anatomically shaped seat ensures that the user is positioned so that he can sit comfortably upright and automatically makes contact with the backrest. The seat must have a soft, rounded front edge to avoid pressure points on the flexing sides of the legs.

Fig. 4.3 Office chair: backrest

- Without pressure on the gluteal muscles, people can sit upright for long. Without pressure on the underside of the thighs, the blood can circulate well in the legs and causes no discomfort.

Upholstery
The task of a swivel chair is to simulate the anatomy of the body and to balance the pressure points. It is breathable and can therefore absorb moisture and release it back into the environment just as quickly. For this reason, the cover must not be glued to the upholstery.

- This significantly increases sitting comfort and ensures concentrated work.

Armrests
The armrests have to fulfil an important function, so that the shoulders are optimally supported and the neck so that neck muscles are relaxed. They must be adjustable

Fig. 4.4 Office chair: armrests

at least in height, better still in height, width and depth, to ensure individual support (Fig. 4.4).

- The relief of the shoulders avoids tension and pain in the neck and neck muscles.

The use of the office chair
The very complex design of the office chair has three major consequences for use in the workplace:

- You should not use it without a briefing and an explanation.
- The adjustment of the chair must be simple, intuitive and comprehensible for the user.
- Depending on the location and duration of use (territorial or non-territorial), you should also consciously avoid using functions in order to avoid incorrect settings.

For the benefit of the physical health of the employee, product design is always at the service of function. Then it is good. Ergonomics is therefore not the antithesis of design but influences this discipline to a great extent. The artist and designer Ray Eames said: "*What works good is better than what looks good, because what works good lasts*" (original quote, from Eames 1981).

Once a person has adjusted his or her personal office chair individually to his or her body measurements and needs, he or she must now devote himself or herself to the relationship to the table and the work equipment on it.

4.2.4   The Table

The worktable should offer sufficient working space depending on the task of the user, so that the work equipment can be arranged individually according to the requirements. The minimum size is 160×80 cm or 1.28 m^2. The table depth of 80 cm has become established since the introduction of flat screens. The height of the work surface must be at least 68 cm to ensure sufficient legroom. In the case of non-adjustable panels, 72 cm is the standard.

As with the task chair, the adjusted height of the table is an important factor that can cause discomfort when sitting at the workplace for long periods (Fig. 4.5).

- If the table is too low, the person often sits with a hunchback on the chair. In addition, it overstretches the neck to retain the line of sight to the screen. It can thus damage the discs in the lumbar and neck area.
- If the table is too high, the person is constantly sitting at the table with his shoulders raised to be able to operate the keyboard and mouse. The result is tension and severe pain in the shoulder and neck area.

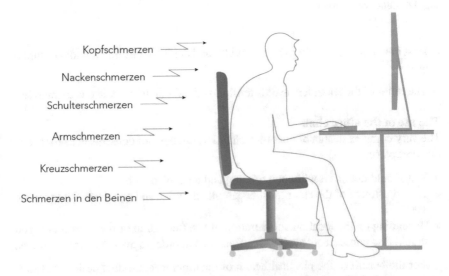

Fig. 4.5 Table too high, no elbow support, the keyboard too far away from the body

The legroom is ensured on the one hand by the minimum height of the table and on the other hand by the basic substructure. The table top must not be free of reflections. This depends on the material, the surface structure and the colour. Glass and metal tables, for example, are completely unsuitable for permanent work. The same applies to very smooth surfaces but also to very dark or very light and brightly coloured surfaces.

A work table with a bright white shiny surface has an effect like a snow-covered glacier in sunlight.

For this reason, there are now special surfaces for the work area to avoid reflections and reflections. Due to a rough surface structure and slightly darkened white tones, white table tops can also be used, for example.

If too bright, light-scattering surfaces, such as a pattern or work table, hit the eyes in the field of vision, this is referred to as *indirect glare*. The adaptive capacity of the eyes is constantly overtaxed. The screen is relatively faint and low in contrast. When working, however, the eyes do not adapt to the low-light screen but to the bright surface, for example, the dazzling work surface. As a result, the eyes are constantly forced to adapt to light and dark, which leads to overstraining the eyes with fatigue symptoms such as burning eyes and tears.

All in all, this is the place: Monotonous sitting damages health.

The opportunity of a height-adjustable table up to the standing height or even a standing desk would enable the screen-oriented employee to alternate between sitting and standing. Not to mention that also long periods of standing should be avoided. The alternation between sitting and standing awakens body and mind.

4.2.5 Screen, Keyboard and Mouse

There are precise recommendations about the arrangement of the work equipment in order to avoid forced postures—but everyone works differently. A single basic rule is usually enough for simplification: Frequently used work equipment should be close to the body (Fig. 4.6). This avoids unnatural stretching movements. The body remains upright in its perpendicular and is thus relieved. Even in the reclined position, the keyboard and mouse can still be operated in a relaxed manner. Elbows can rest on the armrests.

The screen should stand exactly in front of one with a distance of approximately one arm length (to the wrist) (Fig. 4.7). This enables a fatigue-free posture. A slightly lowered view to the screen ensures optimum viewing conditions and prevents neck tension. If the screen is too far away or the font too small, you tend to bend forward. The unnatural posture causes excessive pressure on the neck muscles (Deutsche Gesetzliche Unfallversicherung e. V.). (DGUV) 2015. In order to avoid unpleasant glare to the eyes, the screen should not be positioned either in the direction of the window or in the opposite direction. Therefore, it is always necessary to position the table and the screen at a 90° angle to the window.

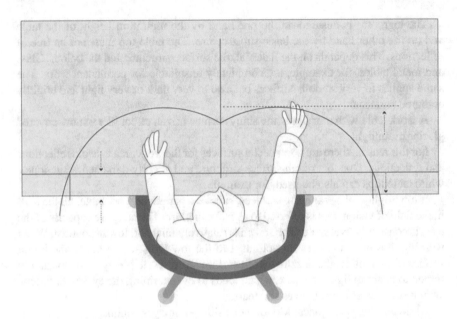

Fig. 4.6 The gripping space on the worktable

The close arrangement of the work equipment and the recommendation of the screen distance also require that the keyboard is freely movable. A distance of approx. 10–15 cm from the front edge of the table is necessary so that the palms of the hands can be placed comfortably. The flat keyboard also allows a natural posture of the wrists. As with the tabletop, a bright version with dark lettering should also be selected for the keyboard. This positive representation, dark writing on a light background, demonstrably ensures better legibility and also avoids annoying contrasts to the screen and table top (ibid.) for the eyes.

RSI syndrome has long been a well-known medical condition. RSI means repetitive strain injury. These are complaints in the neck and shoulder area as well as arm and hand complaints that occur as a result of frequently repetitive activities. RSI syndrome is particularly common during long periods of monotonous work with the keyboard and computer mouse and can range from complaints in the cervical spine and arm to specific diseases such as tendosynovitis or carpal tunnel syndrome.

That is why the correct selection and handling of the mouse should be checked in detail individually and precisely:

- Does the mouse fit the hand in size and shape?
- Is your hand relaxed?
- Can mouse buttons and trackball be operated without cramping your fingers?
- Can it be moved precisely and without large radii on the table (Werner et al. 2012, p. 64)?

Fig. 4.7 The distance to keyboard and monitor

If pain occurs in the shoulder area, the vertical mouse could be an alternative. It enables a relaxed and natural hand posture in contrast to the classic mouse.

This is based on the assumption that "if you let your arms dangle loosely next to your body while standing, they will point inwards towards your body. If you now place your forearms on the table in this same relaxed position, the palms of your hands will also point to each other—your hand will lie vertically on the table" (ibid., p. 66).

All in all, the ergonomic use of work equipment requires a high level of knowledge, interest and commitment on the part of users in order to behave in a way that is good for their health at the workplace. Therefore, perhaps designing a health-promoting environment is more motivating than teaching appropriate behaviour.

4.3 The Office of Options—Activity-Based Working

There is nearly no company that can claim the same operation of work for all employees. In any case, the world of office work is changing increasingly, and many employers are responding to the opportunities of digitalization and the challenges of demographic change with flexible organizational structures and new room concepts. They hope that these will initially lead to improved internal communication and more cooperation. In addition, these so-called multi spaces are often intended to combine a more attractive working environment with significant space reductions. The activity-based working approach is based on these interfaces.

4.3.1 Activity-Based Working—Origin and Explanation

The term activity-based workingwas defined in 1995 in the book "Demise of the office" by the Dutch management consultant Erik Veldhoen (Veldhoen 1995). It originally refers to a concrete concept of work organization and organizational structure. In the meantime, the term is very often used in connection with a special form of office design that provides employees with the optimum workplace for their particular task or activity.

Activity-based working in office furnishing is based on the basic idea that every activity requires an optimal working environment, which is not easily guaranteed by the classic workplace within the framework of any office layout. Within one working day, employees in the office carry out a large number of different "activities": Writing e-mails, telephoning, informal meetings, scheduled meetings, project meetings, research, breaks and more.

In addition, these activities are carried out either alone or in collaboration with other colleagues, and they require varying degrees of concentration. The latter aspect is strongly dependent on the individual personalities. Therefore, the employer offers a variety of different workplaces in the office, which represent the most performance-supporting environment for the most frequently occurring activities in the company.

Activity-based working therefore serves to increase the efficiency and creativity of the employees and can be combined with an attractive and image-promoting working environment. In addition, under certain circumstances, areas can be reduced and thus costs can be saved. Activity-based working is often accompanied by the following factors during implementation:

- Reduction or complete abolition of personally assigned workplaces
- Less traditional workstations in comparison to the number of employees
- Significant increase of workplaces (lounges, project rooms, think tanks, touchdown workstations at benches, etc.) compared to the number of employees
- Implementation or establishment of paperless or paper-reduced offices
- Implementation or establishment of flexible working time and salary models.

4.3.2   Activity-Based Working - Furnishing Aspects

The furnishing of an activity-based office is characterized by its diversity. The individual workplaces require a specific selection of furniture types and—related to the overall appearance of the working environment and, if necessary, taking into account an individual corporate design—an appropriate colour concept (see Chap. 3).

Examples of workspaces in activity-based workspaces and furniture types suitable for them as well as other equipment requirements:

- Classical workstation: Stand-sit table, electrically height-adjustable with height display to simplify adjustment; screens to inhibit sound propagation, absorb sound and protect against visual interference; ergonomic swivel chair with intuitive adjustment options; docking station; keyboard; mouse; monitor holder; (2) screen(s).
- short-term workplace (touchdown workstations): electrified bench solution for up to eight people with acoustically effective screens and wireless charging if required; ergonomic task chairs with intuitive adjustment options.
- Rooms for retreat, plus confidential or concentrated activities, individually or in groups: room-in-room system (think tanks) with individual light and air conditioning control, furniture that is appropriate to the predominant use in terms of function (here several options are often made available in different versions).
- Workshop rooms: flexible tables, foldable and/or on castors and plug and play electrification; flexible chairs (low weight, on castors); smartboards or whiteboards; storage space for workshop materials.
- Creative or project area: Table in standing height; benches or standing aids; smartboard/monitor with wireless connectivity; whiteboard.
- Reception/Homebase: Starting element with full-fledged workplace; sofa; armchair; side tables; storage compartments (locker) can be locked with key or chip; drinks/coffee supply option; e.g. in Scandinavia, monitors are also placed in extensive activity-based workspaces which display the location of individual employees in order to facilitate locating and orientation. Use the lockers as mailboxes.
- Cafeteria/Bistro: Varied, attractive seating at gastronomic level; different table variants (different heights and respective number of seats).
- The following are used for zoning: acoustically effective room divider modules (screens); cupboards for storing remaining documents; converted furniture with a high back and side panels (e.g. alcoves); plants.

4.3.3   Implementation Example

Planning: Vitra (Figs. 4.8 and 4.9).

Map legend
Classical fields of work

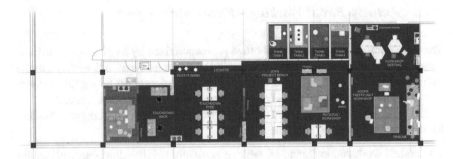

Fig. 4.8 Activity-based working, floor plan

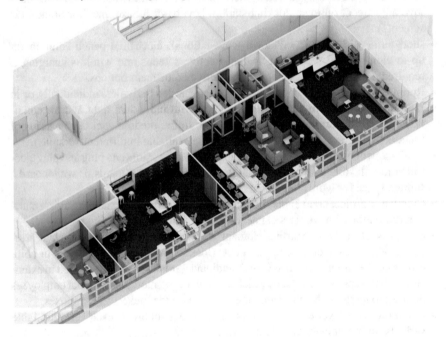

Fig. 4.9 Activity-based working

Computer workstations
Project and team benchmarking
<u>retreat</u>
Think Tank 1: The Blue: meeting with projection and writing surface
Think Tank 2: The Diner: conversations at the table, without technical equipment
Think Tank 3: The Cosy: the relaxed conversation
Think Tank 4: The Cool: concentrated work on the PC alone or in pairs
Touchdown workplace, temporary
<u>Creative and project areas</u>

Post-it wall
Workshop Meeting 2×
Personal storage: Lockers
The Garage
Touchdown workstations, temporary
Agora: meeting place and workshop.

4.3.4 Activity-Based Working (ABW)—Health Relevant Factors

After many years of experience, the activity-based office ("the ABW approach not only encourages people to move around, it can be good for social and collaborative aspects of work", Dowdy 2016, p. 5), which is specifically tailored to the requirements and job profiles, appears to be establishing itself in Germany as well, especially in the Netherlands and Scandinavia. Against the background of the predicted changes in the office work world, many even speak of a disruption, and numerous experts see it as the basis for the office design of the future.

At the same time, according to a study conducted by the job portal Indeed in March 2017 with 1049 respondents, more than 82% of Germans work in cellular offices and more than 81% are very satisfied with this status (Indeed Blog 2017). In addition to the enormous challenges for managers, employees and entire organizations that this process of change will bring (not to mention the psychological complaints that arise in this context); activity-based working in the office requires a new evaluation of the workplace in terms of health protection.

Individual risk assessment becomes the regulating factor. As the Workplace Ordinance does not apply to individual workplaces within the multispace, an individual consideration of the respective work environment is necessary, which the legislator provides in the form of risk assessment. Although, the new Workplace Ordinance in the area of office work has become even clearer: Every employee has the right to a fully equipped computer workstation, regardless of the duration of the work. But this does not necessarily mean that 20 tables must also be available for 20 employees.

In activity-based workspaces the furniture is used by several people. Therefore, they must fulfil the individual physical requirements of all users, the functions must be simplified and the setting options, e.g. for task chairs, must be made more intuitive. Instruction in the various work equipment available is therefore becoming more complex and should be followed up regularly.

Undisputedly, activity-based working provides variety for body and mind through the countless options for different postures and it keeps employees moving. The dynamics in the space result from the type of activities. The change between the working zones automatically leads to a positive effect for muscles and discs.

Another problem is the work of the so-called nomads on mobile devices such as laptops or tablets. Even if the location is changed frequently, the duration of work on the laptop usually does not change. Screen distance and adjustability, flexible keyboards, ergonomic sitting posture are de facto non-existent, the unnatural head posture during tablet use can cause pain in the neck muscles or even damage the discs of the spine after a long time. In this case, the employer has the option of instruction; in this context, the health awareness of the respective managers is also in demand.

Different lighting and colours (s. Chap. 3) also demonstrably improve motivation, concentration and performance. Last but not least, different viewing directions and distances also protect the eyes and eyesight.

It has been proven that the visual connection with nature can improve mental engagement and attention. Blood pressure and heart rate can also be reduced. It also improves the positive attitude and general happiness of employees (cf. Biedermann et al. 2006).

4.4 Plants

In the office, office users often regard them merely as design elements, but in fact they do much more than is expected. The literature shows that a green working environment is permanently more pleasant and healthier for employees and conducive to concentration—and thus also more productive for the company. The mere enrichment of a former Spartan space with plants indeed served to increase productivity and performance by 15% (Nieuwenhuis et al. 2014).

According to this, office workers feel better rested and healthier when they are near plants or windows with a view of greenery (Khan et al. 2005; Fjeld et al. 1998; Kaplan 1993). This is also confirmed by the results of a study in which patients looking out the window at deciduous trees recovered on average one day faster, needed significantly less painkillers and had fewer postoperative complications than patients looking at a stone wall (Ulrich 1993).

Experiments conducted by Fraunhofer were able to fulfil a thesis that windowless offices trigger a general sense of well-being among employees through significantly more plants in the room and a clear view of the plants.

Plants can compensate a missing view to the outside.

Another study found that viewing abstract art increased stress levels by 13%, while nature images reduced stress levels by 3–44% (Salingaros 2012). The participants of a so-called read chip test increased their performance from test to test and also improved their attention capacity, while this did not apply to employees who did not have plants in the office (Raanaas et al. 2010).

Plants or greened walls have the following positive characteristics (see Hahn 2007):

- Optimization of air humidity
- Improvement of room acoustics
- pollutant reduction

The only question that remains to be clarified is who takes care of the maintenance of the plants in the office. The first choice here is a corresponding service provider.

Last but not least, in addition to lighting, colouring and plant design, acoustics make a decisive contribution to keeping employees healthy. In the context of ergonomic office furniture, the following section will therefore focus on acoustically effective furniture.

4.5 Acoustically Effective Furniture

Standards for the room-acoustic planning of multi-person offices can be found in DIN 18041 and VDI 2569. Section 2.3 treated. Although additional sound-absorbing surfaces and the intelligent positioning of acoustically effective furniture cannot replace room acoustics (wall, ceiling, floor), they can achieve a significant reduction in noise levels.

This reduction can be achieved on the one hand through sound refraction : Smooth furniture surfaces promote sound reflection and practically throw the sound back into the room. A perforated surface disturbs the reflection. One part of the sound simply vanishes into the cabinet and disappears, the other part breaks at the edges of the perforation, thereby changing its direction and weakening it.

On the other hand, thicker, soft and open-pored materials absorb the sound and thus reduce the sound energy. This is called sound absorption.

4.5.1 Examples of Acoustically Effective Furniture

A polyester fleece compressed by thermal printing provides an effective surface for damping ambient noise over a large area as a cable trough (Fig. 4.10).

Screens, consisting of pressed, through-dyed polyester fleece or filled with soft sound-absorbing material can provide additional acoustic protection (Fig. 4.11).

Room-in-room systems or sofas with high quilted and padded panels contribute to the visual and acoustic shielding and thus convey a feeling of security, tranquillity and retreat (Fig. 4.12). The acoustic test "Equivalent sound absorption area A in the reverberation chamber according to DIN EN ISO 354—shielding effect based on VDI 3760" is recommended.

Hole pattern fronts and perforated roller shutters break the sound and provide effective acoustic protection in the room (Fig. 4.13).

According to the Workplace Ordinance, 55 dB(A) are prescribed in the office for predominantly mental activities ≤ for all other office work 70 dB(A) (Deutsche Gesetzliche Unfallversicherung e. V.—German Statutory Accident Insurance) ≤70 dB(A) (Deutsche Gesetzliche Unfallversicherung e. V.—German Statutory Accident Insurance) (DGUV) 2015). By the use of suitable materials and surface structures, the sound pressure level can thus be reduced in the room. All in all, the

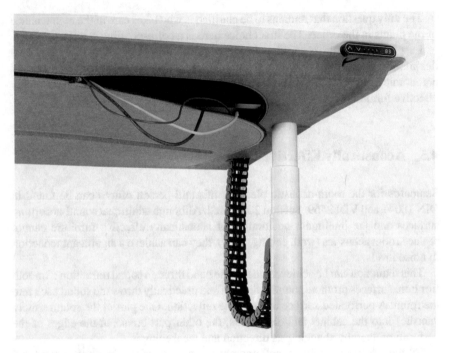

Fig. 4.10 Cable trough

employees feel a calm and pleasant atmosphere, which means that they themselves speak more quietly and mutedly. In addition, acoustically effective lounge furniture creates a quiet retreat zone, which gives the individual the opportunity for concentrated work or offer a small team the opportunity for a meeting without disturbing the environment. All in all, a high quality of stay is created in the room, which has a positive effect on the well-being of the employees.

The health of our employees will be improved due to technological and demographic developments and their impact—from an increase in highly complex job profiles to a shortage of skilled workers up to the probable extension of working life—is becoming an increasingly important factor in the organization and furnishing of offices. Ergonomics continues to play a central role, with the classic concept of ergonomics being extended to other disciplines such as light, air and noise. However, it is even more important to ensure that the organizational structure and the corporate and leadership culture are in line with the constantly changing requirements. Space and adequate office furnishings can support these processes and thus contribute to physical and mental well-being.

Fig. 4.11 Screens

Fig. 4.12 Alcove

Fig. 4.13 Safekeeping

Literature

Bauer, W. (2017). Man in the digitalized world of work. https://blog.iao.fraunhofer.de/der-mensch-in-der-digitalisierten-arbeitswelt/ 6.4.2017. Accessed May 10, 2017.

Biedermann, D., et al. (2006). Pharmacological activities of natural triterpenoids and their therapeutic implications. *Natural Products Reports, 23*(3), 394–411.

Federal Ministry of Labour and Social Affairs. (2016). Green Paper Arbeiten 4.0—Arbeit weiter denken. http://issuu.com/support.bmaspublicispixelpark.de/docs/gruenbuch-arbeiten-vier-null?e=-ProudlyPresents. Access on May 10, 2017.

German statutory accident insurance e. V. (DGUV) (Ed.). (2015). *DGUV Information 215-410 (previously BGI/GUV-I 650)*. Berlin.

Dowdy, C. (2016). Office design that boosts workplace wellbeing. https://www.raconteur.net/business/office-design-that-boosts-workplace-wellbeing. Access 5 October, 2017.

Eames, R. (1981). Interview with Ralph Caplan, February 2, 1981, Herman Miller Archives, Zeeland, Michigan. In: Vitra Design Museum, Library of Congress (Ed.), *The world of Charles & Ray Eames* (p. 21). Berlin: Ernst & Sohn. (1997).

Fjeld, T., Veiersted, B., Sandvik, L., Riise, G., & Levy, F. (1998). The effect of indoor foliage plants on health and discomfort symptoms among office workers. *Indoor and Built Environment, 7,* 204–209.

Hahn, N. V. (2007). "Dry air" and its effects on health—Results of a literature study. *Hazardous Substances-Air Pollution control, 67/3,* 103–107.

Indeed Blog. (2017). *Alone or to several? German office culture today,* http://blog.de.indeed.com/2017/05/22/alleine-oder-zu-zweit/ 22.5.2017. Accessed on ???.

Kaplan, R. (1993). The role of nature in the context of the workplace. *Landscape and Urban Planning, 26,* 193–201.

Khan, A. R., Younis, A., Riaz, A., & Abbas, M. M. (2005). Effects of interior plantscaping on indoor academic environment. *Journal of Agricultural Research, 43,* 235–242.

Nieuwenhuis, M., Knight, C., Postmes, T., & Haslam, S. A. (2014, July 28). The relative benefits of green versus lean office space: Three field experiments. *Journal of Experimental Psychology: Applied. Advance online publication.*

Raanaas, R. K., Horgen Evensen, K., Rich, D., Sjøstrøm, G., & Patil, G. (2010). Benefits of indoor plants on attention capacity in an office setting. *Journal of Environmental Psychology, 31*(2011), 99–105.

Salingaros, N. A. (2012). *Fractal art and architecture reduce physiological stress.* San Antonio: University of Texas at San Antonio, Department of Mathematics San Antonio.

Savills, CCL/Consulting Cum Laude. (2016). *Office of the future? Comparative study on the office of the future from the point of view of Generation X & Y.* http://pdf.euro.savills.co.uk/germany-research/ger-ger-2016/office-of-the-future-de.pdf. Access on May 10, 2017.

Tsunetsugu, Y., et al. (2007). Physiological effects in humans induced by the visual stimulation of room interiors with different wood quantities. *Journal of Wood Science, 53*(1), 11–16.

Ulrich, R. S. (1993). *Hospital gardens turn out to have medical benefits.* Center for Health Systems and Design Colleges of Architecture and Medicine Texas A & M University College State, TX 77843.

Veldhoen, E. (1995). *Demise of the office.* Rotterdam: Uitgeverij 010.

Werner, H., Hansen-Uffenorde, M., Fuchs, B., & Mayer, S. (2012). *Office is in the smallest huts. At home working at desk and PC—Ergonomic and stress-free.* Dortmund: Verlag Dr. Steinert.

Wilke, H. J., Neef, P., Caimi, M., Hoogland, T., & Claes, L. E. (1999). New in vivo measurements of pressures in the intervertebral disc in daily life. *Spine, 24*(8), 755–762.

Chapter 5
Demand-Oriented Building Services Engineering

Plan and Implement Healthier (Administrative) Buildings

Peter Bachmann

5.1 Is Health Measurable? and What Factors Play Important Roles in the Well-Being and Satisfaction of Employees? This Is One of the Central Questions When Designing Office Buildings and Work Environments

From the point of view of the Sentinel House Institute, one of the most important factors is the interior air quality. This medium immediately and directly determines our state of health, our performance and, in the long term, our health. In addition to the question of air exchange in a room, which is treated only marginally, the health quality of the used building materials, equipment materials and furniture regarding the emission of pollutants into the room air, is of great importance for the sojourn quality.

This chapter highlights the fundamental aspects and presents various model projects from several areas.

5.2 Is Healthier Building Worthwhile and What Does It Actually Mean?

Everyone is talking about healthy building, modernising and living. Is this just another gimmick or do real estate companies and their customers benefit from dealing with the topic? The challenges for companies in the real estate industry are enormous.

P. Bachmann (✉)
Freiburg, Germany
e-mail: Bachmann@sentinel-haus.eu

© Springer Nature Switzerland AG 2020
W. Seiferlein and C. Kohlert (eds.), *The Networked Health-Relevant Factors for Office Buildings*, https://doi.org/10.1007/978-3-030-22022-8_5

Government requirements, for example in the area of energy efficiency and sustainability for new construction and renovation, are strict. And almost every day, it becomes more demanding to create affordable living space and to generate an appropriate return at the same time. A term analysis is worthwhile. With the owner builder's increasing awareness of health issues, terms are increasingly used in marketing that are not precise. Here are some hints.

Be careful with the term "healthy". No building material, furniture or building can be healthy. The absolute term "healthy" implies that the product has a healing effect. This is difficult to prove and therefore difficult to assert. Legally, there are a few examples of manufacturers who have suffered severe defeats in court due to advertising "healthy". In this context, the use of the term "healthier" is legally secure and more accurate.

5.3 Biological and Ecological Products and Buildings

Biological products and buildings sound good and enjoy the attention. However, the fact of a biological origin does by no means guarantee a positive health effect. Many biological substances occur in nature, which can have a devastating effect on human health. Venom and plant poisons are ecological and biological but can be fatal and have a negative impact on human health. Some biological products emit solvents (VOCs) into the interior and can trigger allergic reactions in sensitive people.

5.4 Sustainability Must Be Precisely Defined

Sustainability is about consuming resources and goods to the extent that they can be regenerated. Energy efficiency is just one of many topics which the industry is currently concerned with. Are not healthier building, renovating and living a luxury topic under these conditions?

In principle, the current efforts towards sustainability in the construction industry are very welcomed. However, it is just as important that there are clear criteria which provide the user, client and customer with real security and a reference to quality. The question of system boundary quickly arises. Respective assessment systems are not always sufficiently defined in terms of health criteria.

Instead, specific and measurable statements are needed in this context. Marketing and advertising must therefore communicate particularly precisely with regard to health. For example, definite statements can be made about the solvent exposure or formaldehyde in the interior or a product. This statement needs to include the time, the investigation method and the uncertainty of the measurement. The same applies to CO_2 concentrations in an interior. What are the terms of use, what is the exact time of the measurement of the guaranteed CO_2 value?

For products, a precise statement on the pollutant concentration is relatively simple; however, for buildings, it becomes much more complex. Current research results demonstrate the diverse development of results regarding pollutants. Interactions between products, the owner builder's personal contribution, temperature, air humidity and ventilation have an enormous influence on the measurement result. Building providers who have only individual show flats or model buildings measured take the risk of the customer or user being deceived. If the customer really cares about the health quality of an interior is of great importance to the customer, the building must be inspected for harmful substances by an independent measuring institute after new construction or modernisation.

As is so often the case, the evaluation depends on the point of view as well as the background. There are sufficient reasons for healthier construction and the observance of indoor air hygiene.

5.5 Theses for Healthier Building, Renovating and Refurbishing

Unfortunately, pollutants from building products and their processing are still very common. Particularly in the case of renovation, but also in the case of new construction, an airtight building shell for energetic reasons leads to high pollutant concentrations, which can result in health impairments. Against this background, there are many reasons to put the issue of healthier building on the agenda. The following are the ten most important points:

1. Companies can achieve clear legal and liability security with a minimum of effort in terms of pollutant loads and resulting deficiencies. The formulation of contracts with suppliers, architects, specialist planners and craftsmen is crucial. The accurate tendering of health standards for planning and construction services in accordance with VOB is possible, even in the public sector. This has been proven by numerous expert opinions from building law experts. The respective guidelines demonstrate how it works.
2. A concept of healthier building, refurbishing and renovating is part of the company's own quality assurance. According to the principle: "Better informing yourself beforehand, rather than having to expensively repair afterwards with much annoyance and stress".
3. Healthier building does not necessarily have anything to do with ecological but a lot with sustainable constructing. The building products recommended by the partners Sentinel Haus Institut and TÜV Rheinland after thorough testing are well-known brand products that most building and planning departments know and possibly already use. This also means that material costs are not—or at least not significantly—higher than in the past.
4. The criteria on which healthier construction is based are precisely formulated and scientifically defined, for example, by the Federal Environment Agency.

Accredited testing institutes, such as TÜV Rheinland, can test these criteria in accordance with applicable standards and procedures. This allows integration into the sustainability strategy of a construction or real estate company without friction losses.

5. Healthier building can be applied to every construction method and every energetic standard—from terraced houses in the real estate development business to apartment buildings, educational buildings and large administrative buildings.

6. For each phase in which a building is "touched", suitable and cost-optimised concepts are available—from the cosmetic repair after a change of tenant to the basic refurbishment and new construction or a complete core refurbishment. In the latter case, possibly existing pollutants are also examined.

7. Further training of employees and external partners from architecture and trade creates knowledge and experience in and around the company. Once the processes have been put into practice and the material lists have been checked and defined, the employees in the building management and construction department work like before—only with a special focus on the health of the customers and the associated quality of their work. Using the material and knowledge bases, for example, the "Bauverzeichnis Gesündere Gebäude" by Sentinel Haus Institut and TÜV Rheinland, can be helpful, www.bauverzeichnis.gesündere-gebäude.de as it offers valuable first-hand knowledge.

8. Processes are simplified because fundamental questions regarding material selection and processes are clarified. This increases the effectiveness in the company.

9. Through health quality assurance, companies in the real estate industry avoid defects caused by pollutants from building products and their processing, as well as the legal and financial risks that may result.

10. Lastly, health is the greatest good for everyone, naturally also for the customers of the real estate industry. Gaining a positive profile strengthens the company's image in supervisory bodies and the public.

5.6 Quality Seal for Building Materials and Real Estate

One of the most important factors in terms of healthier construction of all types of buildings is the quality of the building materials used. Since not every product can be measured individually for emissions of pollutants, reliable building material labels play an important role. The control of building materials within the framework of German accreditation has been severely restricted by a ruling of the European Court of Justice. The background is the European Construction Products Regulation, which for reasons of competition law does not permit independent national specifications for products harmonised in the EU, i.e. those to which an EU standard applies. The connections are complex, but fundamental, which is why a small excursion is made.

5.7   The ECJ Ruling and Its Consequences for the Construction Industry

Still unnoticed by many players, a ruling by the European Court of Justice on the practice of German building material approval has an enormous impact on all parties involved in the construction process. Architects, in particular, are caught between two stools when it comes to constructing or renovating buildings that meet the criteria of the Federal Environment Agency for healthier interior air. But what is different now?

On 16 October 2016, in Case C-100/13, the verdict of the European Court of Justice (ECJ), which had been handed down two years earlier, came into effect. The EU is thus now also implementing its claim to the priority of the free movement of goods over national regulations in the construction sector. In concrete terms, this means that for construction products harmonised in the EU, i.e. products for which a European standard exists, additional national approval criteria do not apply any more. Accordingly, only construction products with a CE marking may be manufactured, traded and installed for the product group. However, the EU standards on which the CE marking are based only taking into account the conformity of the product with the properties declared by the manufacturer; specifications on emissions of pollutants are not yet included in the EU standards. The European Construction Products Ordinance (BauPVO) expressly stipulates that member states are to work towards supplementing the underlying EU standards. In 2015, Germany had therefore also raised objections to six incompletely harmonised construction product standards, of which the standards for wooden floors and sports floors were rejected. The Federal Government brought an action against this on 19 April 2017. Until the legal dispute has been clarified, the German specifications for emission testing for wooden floors and sports floors will therefore remain valid. For all other product groups relevant to the health quality of indoor spaces, such as floor and wall coverings, adhesives, screeds, wall and ceiling coverings, wood-based materials, plasters, bricks, seals, thermal insulation materials, cements and precast concrete elements, the EU standards on emissions of pollutants do not contain any requirements.

The consequences of the judgement stand in the way of efforts to construct or renovate buildings in such a way that, once they have been completed, they will certainly meet the Federal Environment Agency's criteria for healthier interior air. This also affects buildings with sustainability certifications in accordance with DGNB, BNB, LEED or the NaWoh standard, where excessively high interior loads are defined as "K.o. criteria".

5.8   Responsibility Shifted to Private Level

The ECJ ruling also resulted in a restructuring and amendment of the model building code (MBO). The building regulation lists of the German Institute for Building Technology (DIBt) also lost their validity and were integrated into the new administrative regulation Technical Building Regulations VV TB. Although the fundamental requirements for the safety of buildings have not essentially changed, they are still subject to national regulation. However, responsibilities for the safety of construction products have shifted from the national to the private level—from manufacturers, commerce, architects and construction companies to owner builders and investors. These responsibilities concern all safety-relevant criteria of a building product, in addition to its health quality, for example, also questions of fire safety, stability and sound insulation.

The level of protection to be fulfilled by the owner builder and the commissioned architect remains the same, but the specific requirement "healthy indoor air" was not included in the MBO, contrary to the recommendation of the Federal Environment Agency. It remains to be seen to what extent the federal states will follow the recommendation in their state building regulations. The state building regulations amended by Saxony-Anhalt and North Rhine-Westphalia only contain the previous requirements, for example in § 3 "General requirements".

5.9   A Question of Liability

For investors, construction companies and architects, the current situation is very confusing. On the one hand, when checking product properties, the type of evidence must be clarified. European construction product law provides construction product manufacturers with a multitude of options for declaring the conformity of their products with European requirements. It is the manufacturer's choice whether and what information he provides on the (health) quality of his product. It is also up to the manufacturer whether and in what form he has the health quality of his products monitored by the third parties. Owner builders will try to pass the responsibility on to the construction companies, general contractors or architects. In addition to the responsibility, liability issues would also be passed on accordingly. Architects would thus be responsible for checking declarations for compliance with the legal requirements or the tender submitted by the investor. This leads to not only an enormous administrative effort but also a corresponding liability risk.

5.10   Quality Criteria for Building Material Labels

As a result of the liberalisation of the German construction products market, the significance of test marks has thus increased significantly. Reliable and credible test marks for construction products therefore play a central role. Unfortunately, there are no general binding quality standards. As a user, you have to take a close look at what is tested according to which criteria and how, and what is not. The transparency of the test conditions plays an important role. In principle, it must be ensured that construction products are tested and certified by an accredited laboratory. One example is the TÜV Rheinland test mark "Tested for harmful substances". In cooperation with the Sentinel Haus Institut, this test mark will be placed on the market to create clarity and transparency with regard to the health-related properties of building products. With this demanding mark, manufacturers voluntarily document the high health quality of their building products on the basis of the established AgBB scheme. The validity and contents of the certificate can be checked and examined in the "Certipedia" database of TÜV Rheinland. In addition, there are other alternative test marks for building products which the Sentinel Haus Institut, in cooperation with TÜV Rheinland, recognises as part of its joint initiative "Healthier Buildings" and which are accepted as a requirement for listing building products on the online platform www.bauverzeichnis.gesündere-gebäude.de.

In addition, there is a whole series of other reliable test marks that allow low-emission building material quality. Figure 5.1 shows a small selection in optical form.

Fig. 5.1 Important test marks on the construction products market

5.11   Measuring and Interpreting Pollutants Correctly

Without correct measurements, there is no correct assessment of health quality. Room air measurements are therefore an integral part of health quality assurance in buildings. Their frequency and number are always adapted to the construction project and the type of building. At least one measurement by an independent expert up to four weeks after completion of the building is mandatory. For investors, planners as well as for private and public clients, the handling of pollutant measurements in room air, dust or materials is a delicate matter. A lot depends on the results: Does the building have to be closed completely or partially? Is a renovation necessary and how expensive will it be? What are the health risks to children and staff? What reactions do parents and the media show? And last but not least: Has the agreed room air quality actually been achieved in a new building or renovation? The examples show that only a valid, reproducible measurement strategy provides a reliable data basis.

The requirements for interior investigations are defined: The contractor should have both a system certification according to the DIN EN ISO 9000 ff. series of standards and a laboratory accreditation according to DIN EN ISO/IEC 17025. This proves that employees undergo regular further training that only tested, and suitable measuring instruments are used and that the testing laboratory regularly participates in interlaboratory surveys for quality assurance.

If cheap equipment or non-recognised methods are applied, the results will not withstand a later, versed examination or lawsuit. A comparison with other measurements or official values is hardly possible. Investigations commissioned for a lot of money can turn out to be worthless. Also, only measurements for which evaluation criteria exist should be carried out. It is important that the boundary conditions prevailing during sampling are recorded and documented in the result report. Temperature, relative humidity and ventilation behaviour before and during room air tests have a major influence on the test result.

Before measuring, a coordination meeting should take place between the client and the expert, the result of which should lead to a task-oriented study plan in accordance with VDI 4300 Part 1. An experienced and qualified expert is also indispensable for the interpretation of the results. He is also able to identify the responsible sources of pollutants in the case of conspicuous findings and to recommend and communicate measures that lead to an improvement of the situation.

5.12   Refurbishment Activates Contaminated Sites

Further risk factors for the health of building users lie in existing buildings, to which only a brief reference is made. In addition to mould and moisture damage caused by construction defects, contaminated sites such as asbestos, PCB, wood preservatives or tar-containing adhesives are further risk factors for the health. "If these were inconspicuous in the past or were not identified as the cause of health problems, they

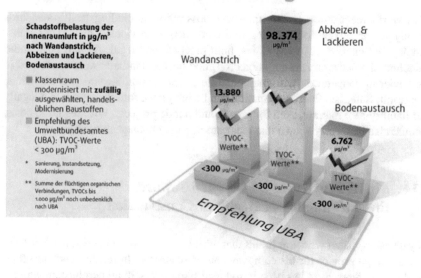

Fig. 5.2   Measurement results after typical refurbishment cycles

can be uncovered by the opening of components in the course of a renovation and thus released and activated", emphasises Dr. Walter Dormagen, Division Manager at TÜV Rheinland. In view of the risks and the large number of possible product combinations, a renovation or refurbishment without accompanying quality assurance is like a blind flight. This also applies to the topic of ventilation. Carbon dioxide is produced during breathing and can impair attention and the ability to concentrate even in comparatively small quantities. It is necessary to check the correct implementation of the ventilation concept and the accurate installation and maintenance of a ventilation system.

Newly installed building products can also significantly impair indoor air quality in buildings in the course of renovation or refurbishment, as Fig. 5.2 shows. Those who do not order pollutant-tested products must expect enormous concentrations of volatile organic compounds (VOC), as the measurement results after typical remediation cycles show. The Federal Environment Agency recommends a long-term target value of less than 300 μg/m^3 room air. Up to 1000 μg/m^3 are still harmless, from 3000 μg/m^3 the use is acceptable for a maximum of one month. From 25,000 μg/m^3 the hygiene situation is unacceptable, and measures must be taken. It is the task of facility management to plan the work accordingly and to put it out to tender so that materials and processing do not cause any permanent health problems.

5.13   Costs and Award

Also worth mentioning is the important possibility of binding health standards already in the tender for planning and construction services. This creates transparency and clarifies responsibilities, both practical and legal. As part of a tender guideline, the necessary formulations for good health standards were developed in the tender in cooperation with municipalities, construction lawyers and engineers. The result: Safe, healthier construction can be integrated into the tender in a legally and liability-safe manner, both for private and public projects. With regard to costs, it could also be clearly shown that these are in a very manageable range.

5.14   Model Projects as a First Step Towards the Introduction of Healthier Construction

Any decision-maker who now thinks that the introduction of safe, healthier construction is a huge project for his company can be reassured. In recent years, healthier building and renovation has been introduced step by step in numerous companies.

From strategic consulting for the management to training for in-house and external planners to training for craftsmen, the implementation complies with the wishes and possibilities of the company. On closer inspection, many managers in the real estate industry will find that their company is already on the right track and that the steps towards healthier construction and renovation are small.

A very practical possibility is a model project. One example is the construction of a new day care centre in Düsseldorf's Kuthsweg by Rheinwohnungsbau GmbH. The materials used were checked for the client. Two training sessions for the craftsmen involved were conducted by CO architects in Solingen, trained specialist planners for healthier construction, on behalf of the Sentinel Haus Institute. The support of the client and the architects was provided by a residential health coordinator, "WoGeKo" for short, who controlled the compliance with the agreed construction site rules during construction site appointments and served as a contact person. For the day care centre Kuthsweg, this role was filled by CO architects on behalf of the company. The first meeting took place in June before the screed was laid, further meetings followed in the course of further expansion work. Finally, the day care centre was certified according to the criteria of the Sentinel Health Passport. Buildings are currently being awarded a joint test certificate by TÜV Rheinland and Sentinel Haus Institute (Fig. 5.3).

A further possibility to gain knowledge about construction methods and the behaviour of used products are model rooms. Here, too, TÜV Rheinland and the Sentinel Haus Institut have gained a wealth of knowledge in the field of educational institutions, which can of course be applied or transferred to other types of buildings.

The model project "Healthy school living space" (Fig. 5.4) shows in practice how the healthier building works. For this purpose, two model rooms were set up

Fig. 5.3 Day care centre Kuthsweg in Düsseldorf is integrated into a housing project of social housing and certified according to the high health standards of the Sentinel Haus Institute

on the TÜV Rheinland site, one with specifically selected materials tested for their health suitability, the other with randomly purchased materials. A measurement programme recorded volatile organic substances (VOC—cf. 31. BImSchV (BGBl. I No. 44/2001); ChemVOC-FarbV (BGBl. I 70/2004)—and SVOC) and formaldehyde as key parameters. On this basis, several renewal cycles were also simulated, as is usual in the renovation and refurbishment of schools and day care centres. The influence of furniture and the indoor air pollution of cleaning agents were also measured.

The measurements show clear differences between health tested and untested products. Not least in the case of renovation and refurbishment work with tight deadlines, the use of untested products resulted in some cases in massive exceedances of the recommended values. In some cases, according to the recommendations of the Federal Environment Agency, the use would have been limited in time. The experiences of TÜV Rheinland and the Sentinel Haus Institute regularly show abnormalities with regard to health and hygienic quality in newly built or renovated educational facilities. The most frequent triggers are solvents, formaldehyde, plasticisers and so-called pot preservatives, which can trigger or intensify allergies. In contrast, in the newly erected room with the tested materials, the strict guideline values of the Federal Environment Agency were undercut after only seven days.

In the model project, it became clear that good indoor air is not a coincidence but the result of consistent quality assurance. A list of the construction products selected for the tested healthier model room and other model projects can be found

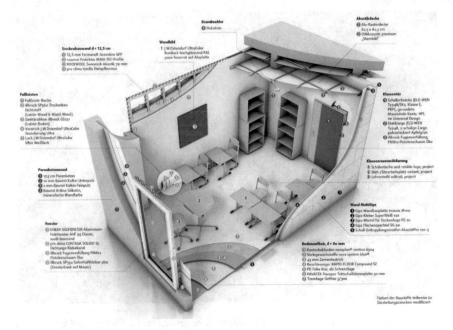

Fig. 5.4 Model project "Healthy living space school" (German)

in the "Healthier Buildings" database, www.bauverzeichnis.gesündere-gebäude.de.
General Information: www.tuv.com/gesundes-bauen-projekt.

5.15 "My Future Office" Research Project

On the initiative of the window and façade manufacturer SCHÜCO and other international building material and system suppliers, a research project is currently being implemented. MY FUTURE OFFICE is managed by the Sentinel Haus Institute and realised in cooperation with TÜV Rheinland. It combines the expertise of experts and numerous manufacturers. The aim of the project, which is scheduled to run for several years, is to achieve knowledge and standards for healthier, performance-enhancing and profitable office buildings. The reasons for this initiative are obvious:

- Modern energy-efficient buildings with airtight building shells need good construction quality management in order to make the healthier properties affordable, plannable and practical.
- In the international competition for the best employees, an optimal modern working environment helps. The emotional theme of health and well-being positively enhances the value of the property and binds employees to the company.

- Promotion of performance reduction of pollutants (CO_2, solvents, formaldehyde, etc.).
- Minimisation of downtime, reduction of pathogens and risk of infection.
- Motivating work environment due to optimal lighting conditions, temperature, acoustics, etc.
- Increasing the value of the real estate through applied sustainability.

Within the framework of the project, building material and material system are tested for harmful substances. In addition, a model room is equipped with different techniques and materials and their effects on the health situation are examined and evaluated (see www.my-future-office.de).

Chapter 6
Medical Aspects

Christian K. Lackner and Karin Burghofer

6.1 Sitting in the Office

In principle, permanent sitting is by far not suitable for the human musculoskeletal system because the muscles are forced to carry out predominantly ongoing static holding work—which, by the way, also applies to prolonged standing in one place.

Interestingly, occupational physiological studies have shown that static muscle work—compared to more dynamic work—leads to a higher heart rate and significant longer recovery intervals.

Through so-called dynamic sitting manner means, that by frequently changing the sitting position—sometimes upright or slightly leaning back, sometimes leaning forward—or ideally by repeatedly switching between sitting, standing and walking, the proportion of negative static holding work is significantly reduced (Bundesanstalt für Arbeitsschutz und Arbeitsmedizin (BAuA) 2016).

Measurements of intravertebral disc pressure bearing load on the lumbar spine (lumbar spine) have shown that the physiologically best sitting posture is a slightly reclined sitting position ($\rightarrow$ large body opening angle = lower internal disc pressure).

"Dynamic" sitting, i.e. the repeated alternation between leaning forward and rather upright sitting or the slightly reclined sitting position, induces the physiologically desired metabolic activity of the lumbar spine intervertebral discs and is therefore gentle on the back and therefore a good prophylaxis against so-called back pain. "Back pain".

C. K. Lackner (✉)
Munich, Germany
e-mail: christian.lackner@dreso.com

K. Burghofer
Bad Tölz, Germany
e-mail: praxis@burghofer.de

© Springer Nature Switzerland AG 2020
W. Seiferlein and C. Kohlert (eds.), *The Networked Health-Relevant Factors for Office Buildings*, https://doi.org/10.1007/978-3-030-22022-8_6

The seat surface of modern office chairs—often called "orthopaedic" or "anatomically"—is shaped in such a specific way that the effective weight of the seated person is optimally absorbed while a uniform distribution of pressure is guaranteed.

In terms of stability and steadiness, such modern office work chairs are designed in a way that they can provide the choice of various safe seating positions—so-called dynamic sitting.

The above-mentioned aspects also apply analogously to the so-called knee-chairs, ball chairs and sprung stools. However, sitting all day without a backrest is clearly a strain on the intervertebral discs in the lumbar region and should therefore be avoided (Accident Insurance 2008).

6.2 Sitting or Standing

In 2013, the Harvard Business Review published an overview article on the effects of sitting in the workplace. For those working in the office, the "sitting at work" time interval has an average volume of 9.3 h per working day. In comparison, daily sleep in this population averages only 7.7 h (Merchant 2013).

Sitting in the office is so ubiquitous that we often no longer ask ourselves how often and for how long we do it. A large number of health studies have concluded that people should generally sit less, get up more and move more. Already after sitting for one hour, the body's own production of enzymes, which are decisive for fat burning in the metabolism, decreases measurably by up to 90%. Extended sitting slows down the body's metabolism, which also influences the HDL level (so-called good cholesterol) in our body. Findings from research show that this lack of physical activity is associated with 6% for heart disease, 7% for type-2-diabetes, and 6% for other diseases and 10% is associated with breast cancer or colon cancer (ibid.).

There is a large number of findings in the literature that show that sitting too long entails health disadvantages. Researchers have already found the sitting posture with hypertension diabetes, obesity and arteriosclerosis were connected.

The so-called standing tables, also apostrophized as "diseases of civilization" , are more often postulated by manufacturers' marketing, but there is still a lack of really valid and reproducible evidence. Nevertheless, in recent years we have experienced a kind of mainstreaming of the standing table in the office. This is probably a fundamental step in the right direction—from a medical point of view, however, the intermittent change between sitting and standing position will only predominantly induce positive orthopaedic effects—in contrast to climbing stairs instead of using lifts.

Technical innovations in everyday office life also have medical effects. The introduction of small, mobile and lightweight notebooks with flat screens means that the fixed connection between screen and keyboard means that they can no longer be arranged flexibly on the desk. This also makes it more difficult to react to the lighting conditions and reflections in the office.

Hereby, anti-reflective displays and correct positioning do play an outstanding role in avoiding unergonomic sitting postures. In addition, notebook keyboards are usually smaller than standard desktop keyboards. These aspects can be very effectively counteracted in the modern working environment by docking stations and separated flat screens as well as connected keyboard and mouse.

6.3 Green Plants in the Office

British researchers around Nieuwenhuis and Knight (2014) found in 2014 that an new offices from the first are designed with green plants makes employees significantly more and deeper satisfied and that they are 15% more willing to perform than people at a workplace without plants (see checklist, Sect. 10.1).

According to the scientists, this is the first study of its kind in a real office setting. The positive influence of green plants on our work performance has already been evaluated more often under laboratory conditions (in vitro). For the first time, however, the results of this particular study were obtained directly from the real and well-known office environment of the study population.

The green workplace was defined in such a way that at least two green plants were visible to the employees from each workplace. Parallel to the remodelling, the researchers asked the employees how they perceive the spatial working atmosphere. In addition, the researchers analysed the productivity of the personnel over the entire trial interval.

The population examined here felt measurably and significantly more comfortable in an office designed with green plants was able to concentrate better on their duties/work according to their own statements and also benefited from the measurably improved ambient air. The results of the study showed that the cohort that stayed in a "green office" felt more physically, mentally and emotionally connected to their office workplace than their peer group (comparing cohort). The subjective well-being was significantly better and so was their ability to concentrate.

Also the quality of the air to breathe is measurably better in green offices. By subjectively perceived "bad" air, we generally mean ambient air with an increased carbon dioxide and lower oxygen content. At workplaces in rooms with a lower oxygen content, physical and mental performance continuously decrease measurably. Plants convert the exhaled carbon dioxide into oxygen and thus improve the air composition.

All these factors together resulted in both—the motivation and the performance of the employees being higher in the green office than in the non-plant-based offices. The green workplace environment not only had a positive effect on the health and well-being of the employees, but also on the turnover of the company. Not only did these people feel subjectively much better in workplaces, they were objectively more creative, productive and also more willing to perform when they had green plants in their direct environment (Korpela and De Bloom 2017).

In a large survey in England with 7000 office workers based on this survey, it was found that 40% of office workers had no green plants in their office environment and their workplaces at the time of the survey. In addition, almost 30% had actual no natural light source at their office workplace (Nieuwenhuis 2014).

In this study, green plants in the office resulted in significant a lower pulse rate and a lower increase in blood pressure when stress was felt.

In the collective examined here, it was noticeable that the inflammation values in particular were significantly lower in the accompanying laboratory parameters and thus the tendency and vulnerability to infectious diseases and also cardiovascular complaints was less pronounced (Nieuwenhuis et al. 2014; An and Colarelli 2016).

It is therefore healthier and more gentle for the human organism if it perceives plants around it and finds itself in a green workplace environment.

These findings are consistent with examinations of hospital patients: The healing process is faster and patients recover more quickly from surgery if they look out of their patient room on the ward into a park instead of a parking lot (Ulrich 1984).

These are essential indicators that simple methods—such as the greening/revegetation of workplaces or hospital areas—lead to people staying there not only being in a subjectively more pleasant environment, but also being objectively more resistant to infections and thus contributing to their health (Theodore 2016).

Already older studies have shown that the sight of plants reduces stress and increases the attention span and well-being (Dravigne et al. 2008). For the first time, the British studies were able to measure tangible how office plants affect productivity at the workplace. "Employers should be able to keep their penchant for a sober office s", emphasizes Nieuwenhuis, "not only for the benefit of the employees, but also with regard to the success of the company" (Nieuwenhuis et al. 2014).

6.4 Psyche and Office Work

Every workplace exerts implicit influences on people working there. Psychological stress can therefore also be inseparably linked to work at an office workplace (Fig. 6.1).

Psychological stress has always been an existing component of (office) work (Berger 2012). However, stress in the office can lead to a considerable impairment of well-being and subjective quality of life. lead. In a survey on the extent to which works of art can reduce stress, 13% of respondents answered "Yes". In comparison with nature scenes were shown to the co-workers. According to Salingaros (2012), biophilic designed art reduces physical stress; stress processing: people prefer the dimensions space, nature, rich in species, refuge, culture, perspective and social aspects. The dimensions of shelter and nature are most strongly correlated with stress, indicating the need to find the most restorative environments. Urban green spaces help with stress management.

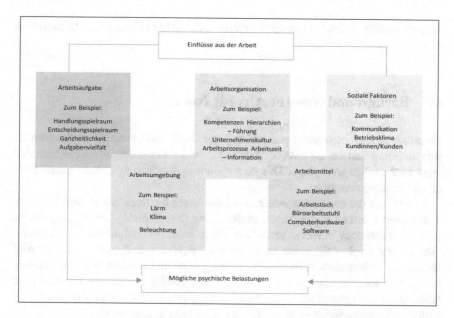

Fig. 6.1 Influences of office work on people and possible psychological stress, according to DGUV (2016), pp. 215–410

On the other hand, red objects in red rooms show a higher stress level than green or white room conditions. Consequently, the research proves that environmental colours play a significant role in stress perception (cf. Kutchma 2003).

Therefore, stress management must be successfully and sustainably anchored in the corporate culture (Lohmann-Haislah 2012).

A successful strategy is a lastingly anchored stress prevention culture—this is anti-stress-culture strategically combines successful and healthy work in the company. This is essentially about promoting health-conscious behaviour among office employees and their managers (Schweer und Kummreich 2009).

Office workers in middle and upper management positions suffer in particular from a perceived lack of influence in their own work context from the point of view of the employee of existing contradictory institutional requirements and a lack of recognition of their own performance by the company, their peers and its line managers (ibid.).

Therefore, it is important to develop emotional and social competence in order to in these aspects—especially in middle and upper management. Through specific training measures for managers, known risk factors for burnout mechanisms in the office can be effectively reduced at an early stage and first signs can be identified and treated.

This would make it possible to identify typical trigger mechanisms for negative stress reactions such as (too) high work demands with inherent role conflicts and continued pressure from superiors/individual line management and a resulting deteriorated working climate in an earlier phase and to effectively combat the first signs of this.

6.5 Background Noise or Already Noise

Even though subjective noise perception is very individual, there are sufficient findings that show that a noise level of 55 dB(A) should not be exceeded for concentrated work in the office workplace. This corresponds approximately to the volume level of a normal conversation.

The higher the need for concentration, the lower the threshold at which sounds and/or conversations are perceived as disturbing [35–35 dB(A)]. Here, innovations in office technology in recent years have led to noise immissions being reduced by The current level of the noise reduction was effectively limited to a maximum of 48 dB(A). This mainly is in relation of computers and printers and there mainly fans and blowers for cooling the housings (Evans and Johnson 2000). The zoning of office space workplaces with designated activity profiles and thus noise profiles up to noise-absorbing furniture also takes these findings into account.

An interesting aspect is the increasing use of musical background sound in larger office spaces. On the one hand, music can provide relaxation, but on the other it can also induce stress. Depending on the respective employee (emotion/mood, life circumstances and age), music is able to reduce internal tensions and strengthen concentration and performance at the workplace (Bernardi et al. 2009).

Quiet background music —for relaxation and concentration —occupies an ever-increasing space in the modern office environment. In medicine, music has proven effective in lowering blood pressure and preventing stress, for example. Here instrumental music is clearly preferred, since music with singing can have a stress-inducing effect. Music works with many abrupt changes, and jumps in rhythm are also critical and in the volume .

The key of a piece of music contributes significantly to the fact that a person's mood can change when listening to this music. Melancholic tones create a contemplative, sometimes even sad or melancholic mood. Major keys are associated with a cheerful, uplifting mood. In spite of these certainly correct observations, each piece of music has an individual effect (Trappe 2009).

However, it has been scientifically proven that quiet background music not only significantly promotes sales, but also improves the concentration and performance of employees with the right selection, volume and dosage.

6.6 Hygiene in the Workplace

Hygiene in the workplace is basically and primarily infection prevention . It is thus based on complex knowledge about the origin of infections and is thus of a multilayered nature and anything but simple.

Hygiene in the office and at the workplace also affects the floors in particular. Office floor coverings are intended dirt traps and therefore become a hygiene problem consecutively. Therefore, office floor coverings (carpets, linoleum or PVC floors) must be cleaned regularly and thoroughly /to be cared for in order to be able to continue to guarantee hygiene in the office.

Carpets in the office increase the risk of dirt and dust deposits. Commercial carpet surfaces must be treated regularly with carpet cleaning equipment and disinfectants. Specialist firm are cleaning companies shampoo carpets with spray extractors and use special carpet cleaners that ensure cleanliness and durability.

Hospital evidence and clinical findings show clearly that hygiene problems can occur more quickly with linoleum or PVC floor coverings, as the polyurethane coating is often only a few micrometres thin and is easily attacked by dirt particles. In the resulting fine cracks or microscopically small holes , viruses and bacteria settle within a very short period of time.

Lots of findings in larger hospitals show clearly that rubber floor coverings offer a more hygienic alternative here, as their extremely dense surface and factory UV cross-linking mean that they require neither coating nor varnishing.

For waterproof floors, cleaning the floor has become a with pH-neutral wipe care hygienically proven, which prevents the formation of layers and streaks. However, so-called pore-deep floor cleaning and maintenance also includes the removal of older residues of cleaning agents and subsequently protects the floor covering from re-contamination.

When furnishing shower rooms, it should be noted that these are subject to the effective federal workplace regulations (ArbStättV 4.1 para. 2). These regulations and requirements are concretized by the technical regulations (ASR for sanitation rooms). Here, the essential requirements for the operation of shower and washrooms in buildings are formulated.

In addition you have to carry out mandatory Legionella prophylaxis of the water strands, these hygiene requirements for the rooms are . This is certainly an important decision and clarification aspect for or against the furnishing of shower rooms in office buildings.

The subsequent establishment of shower facilities usually only makes economic sense in the immediate vicinity of existing toilet facilities due to the technical building infrastructural requirements (zoning). However, as a rule the equally necessary and associated changing facilities are lacking here. This is often the reason why shower facilities within office buildings are moved to the basement.

When it comes to hygiene in modern office buildings, it goes without saying that a critical look is also taken at the ventilation and air-conditioning (HVAC) systems.

Scientific findings on the health benefits or damage of such technical-venting-systems originate primarily from hospital hygiene.

In the literature, there are several indications that postoperative wound infection are induced by air-conditioning systems and are more likely to cause air pollution than to prevent it. Construction, maintenance and servicing of air-conditioning systems are cost-intensive; failure to carry out regular maintenance is the classical root cause of nosocomial infections. However, it also becomes clear from the contributions that air is of secondary importance as a pathogen reservoir for postoperative wound infections compared to the behaviour of the surgical team/ward personnel (Just 2007).

The importance of hand hygiene in the professional environment is identified from a large number of findings in numerous publications. The mechanisms are clearly identified and described, and methods for effective remedial action are available. The only thing missing is consistent implementation and continuous application, which is equally pronounced in medicine and the office workplace environment (Bergler 2009).

That brings us back to the staff. Basically, taking a meal at the workplace is not really a good idea and a considerable source of germs at your own desk for keyboards, computer mice and telephones. Hygiene at the workplace can hardly be maintained in this way—germs and bacteria can thus be found widely scattered at the office workplace.

Food leftovers or spilled drinks thus promote germ growth, eating at the desk becomes a hygiene trap and must be avoided wherever possible.

Studies show that the transmission of pathogenic germs, especially in the winter months in open-plan offices and with desk-sharing-policies in the office, should not be underestimated. Individually, packaged disinfectant wipes are recommended and very easy to use for hygienic disinfection. of hands, tools and smaller areas. In combination with regular hand washing, these disinfectant wipes protect against the spread of infections and prevent fungal and bacterial infections—a simple but very effective measure for more hygiene in the workplace.

More often computer keyboards were cited as a possible source of germ transmission in the field of workplaces. In the meantime, computer keyboards have been developed that are coated with antimicrobial polymers (often Biosafe HM 4100), which have led to a demonstrable and significant reduction in the germ transmission rate between users in clinical and office environments (D'Antonio et al. 2013; Hartmann et al. 2014). However, this effect has generally not yet been demonstrated for (hospital) furniture surfaces.

Hereby, it can be seen that manufacturers of office furniture and components for open-plan offices are increasingly advertising their ability to offer surfaces with "antimicrobial equipment". Having failed to prove its usefulness in high-stress environments such as high-risk clinical areas, modern office environments have now been identified as a potential profitable customer target group that may not or not yet be aware of the current state of science in this context.

No reliable literature data exists so far, neither in laboratory tests (in-votro) nor in reality (in vivo), that a reduction of transmissible infections or germs can actually and reproducibly be achieved on surfaces equipped in this way. Even the reduction factors allegedly measured in the laboratory can be doubted with regard to their level, since as a rule it is not the number of re-isolatable microorganisms that was taken as the initial germ count, but only the number in the suspension applied, which only provides a higher germ killing (Wille 2013).

In 2012, Charles Gerba and his colleagues from the University of Arizona conducted one of the largest surveys of surfaces in office buildings. Over 5000 surfaces in law firms, insurance companies, call centres and companies in the healthcare sector were examined for their germ contamination. At the top of the list of surfaces contaminated with pathogenic germs in office environments are the taps in office kitchens with 75% alarming concentrations of germs, followed by microwave doors (48%) and refrigerator handles (26%) (Harrison and Gerba 2012).

In a follow-up study that followed, Kelly Reynolds and Gerba were able to scientifically prove that employees who come to work with infectious diseases (predominantly banal viral diseases) distributed their pathogens on the mainly shared surfaces in the office environment by lunchtime.

After the employees were instructed about banal measures of hand and surface hygiene and these were also checked in the following investigation period, the infestation of further germs could be reduced by 80% (Reynolds et al. 2013).

This illustrates once again that hygiene is largely an issue of personnel discipline and cannot be delegated to technological devices/preparations. As annoying as that may seem.

References

Accident Insurance, I. F. (2008). *Ergonomic examination of special office chairs*. BGIA Report 5 Sankt Augustin: Institute for Occupational Safety and Health of the German Statutory Accident Insurance.

An, M., & Colarelli, S. (2016). Why we need more nature at work: Effects of natural elements and sunlight on mental health and work attitudes. *PLoS ONE, 23*(5), 1–17.

Bergler, R. (2009). *Psychology of hygiene*. Frankfurt/M.: Steinkopff Publishers.

Berger, M. (2012). *Position paper of the German society of psychiatry, psychotherapy and neurology (DGPPN) on burnout*. Berlin: DGPPN.

Bernardi, L., Porta, C., & Casucci, G. (2009). Dynamic interactions between musical, cardiovascular, and cerebral rhythms in humans. *Circulation, 119*(30), 3171–3180.

Dguv, D. G. (2016). *Monitor and office workstations—Guide in DGUV-Information 215–410*. Munich: German Statutory Accident Insurance.

D'Antonio, N., Rihs, J., & Strout, J. (2013). Computer keyboard covers impregnated with a novel antimicrobial polymer significantly reduce microbial contamination. *American Journal of Infection Control, 41*, 337–339.

Dravigne, A., Waliczek, T., & Lineberger, R. (2008). The effects of live plants and window ews of green spaces on employee perceptions of job satisfaction. *HOARD, 43*, 183–187.

Evans, G., & Johnson, D. (2000). Stress and open-office-noise. *Journal for Applied Psychology, 85*(5), 779–783.

Federal Institute for Occupational Safety and Health (BAuA). (2016). *Up and down—Again and again—More health in the office through sit-stand dynamics*. Dortmund: BAuA.

Grahn, Patrick, & Stigsdotter, Ulrika K. (2010). The relation between perceived sensory dimensions of urban green space and stress restoration. *Landscape and Urban Planning, 94*(3-4), 264–275.

Harrison, J., & Gerba, C. (2012, May 23). *Where the germs are: Office kitchens, break rooms.* University of Arizona. https://uanews.arizona.edu/story/where-the-germs-are-office-kitchens-break-rooms. Accessed 10 May 2017.

Hartmann, B., Benson, M., & Junger, A. (2014). Computer keyboard and mouse as a reservoir of pathogens in ICU. *Journal of Clinical Monitoring and Computing, 18*, 7–12.

Just, H.-M. (2007). Infection prevention or hygiene? *Hospital Hygiene, 2*(2), 89–91.

Korpela, K., & De Bloom, J. (2017). Nature at work: Links between window views, indoor plants, outdoor activities and employee well-being. *Landscape and Urban Planning, 160,* 38–47.

Kutchma, T. M. (2003). The effects of room color on stress perception: Red versus green environments. *Journal of Undergraduate Research at Minnesota State University, Mankato, 3,* Article 3.

Lohmann-Haislah, A. (2012). *Stressreport Deutschland 2012—Psychological requirements, resources and well-being.* Berlin: BAuA.

Merchant, N. (2013, January 14). Sitting is the smoking of our generation. *Harvard Business Review.* https://hbr.org/2013/01/sitting-is-the-smoking-of-our-generation. Accessed 10 May 2017.

Nieuwenhuis, M. (2014, September 1). Why plants in the office make us more productive. University of Exeter. http://www.exeter.ac.uk/news/featurednews/title. Accessed 10 May 2017.

Nieuwenhuis, M., Knight, C., Postmes, T., & Haslam, S. A. (2014, January 1). The relative benefits of green versus lean office space: Three field experiments. *Journal of Experimental Psychology.* http://dx.doi.org/10.1037/xap0000024. Accessed 10 May 2017.

Petersen, J. (2006). Computer workstations—An occupational health evaluation. *Deutsches Ärzteblatt, 103*(30), 1704–1709.

Reynolds, K., Gerba, C., & Blue, A. (2013, January 1). *Germs spread fast at work, study finds.* From: University of Arizona. https://uanews.arizona.edu/story/germs-spread-fast-at-work-study-finds. Accessed 10 May 2017.

Salingaros, N. A. (2012). *Fractal art and architecture reduce physiological stress.* University of Texas at San Antonio, Department of Mathematics San Antonio, TX 78249 U.S.A. yxk833@my.utsa.edu.

Schweer, R., & Kummreich, U. (2009). Health competence and prevention culture-indicators for health and success in companies: A practical model for action. *Journal of Ergonomics, 63,* 293–302.

Theodore, D. (2016). Better design, better hospitals. *CMAJ, 188*(12), 902–903.

Trappe, H.-J. (2009). Music and health. *German Medical Weekly, 134,* 2601–2606.

Ulrich, R. (1984). View through a window may influence recovery from surgery. *Science, 224*(4647), 420–421.

Will, B. (2013). Antibacterial equipment and surfaces. *Hospital Hygiene and Infection Prevention, 35*(5), 155–160.

Chapter 7
Physical Activity in the Modern Working World

Michael Christmann

7.1 Preliminary Remarks

The right amount of exercise is healthy for both the healthy and the sick. This has meanwhile been proven beyond any doubt (cf. Pagenstert 2017). The times when people suffering from cancer, heart or lung disease were advised to rest as much as possible should be over, even if this has not yet become a matter of course for every single doctor.

Somewhat less widespread is the knowledge that there is no need for athletic performance in the narrow sense to prevent a wide range of illnesses, including depression and dementia but that sufficient everyday activity also fulfils the purpose.

However, this knowledge is not new either. As early as 1953, a study was published in which an increased cardiovascular risk of bus drivers compared to bus conductors was described impressively (cf. Morris et al. 1953), although it was not possible at that time to prove a causal relationship.

Thomas Jefferson, one of the founding fathers of the USA, was already familiar with the connection between movement and health as non-scientific implicit popular knowledge or as intuitive knowledge; "Give about 2 h every day to exercise, for health must not be sacrificed to learning" (Boyd et al. 1950). It is not surprising that similar statements are found in Hippocrates and other ancient authors. In old Ayurvedic texts from India, even the positive effect of physical activity on mental health is mentioned (cf. Tipton 2008).

In the first systematic scientific studies on the relationship between physical activity and health, the focus was on everyday activities that led to an increase in energy turnover—i.e. working people with movement at work compared with those who worked primarily sitting (cf. Morris et al. 1953; Paffenbarger et al. 1986).

M. Christmann (✉)
Sanofi-Aventis Germany GmbH, Frankfurt am Main, Germany
e-mail: Michael.Christmann@sanofi.com

© Springer Nature Switzerland AG 2020
W. Seiferlein and C. Kohlert (eds.), *The Networked Health-Relevant Factors for Office Buildings*, https://doi.org/10.1007/978-3-030-22022-8_7

Since technical development led to the fact that less and less physical activity was necessary in the performance of work and generally in everyday life and in addition the sitting times during leisure time increased, the daily scope of movement in the average population decreased continuously. Hence, the research focus shifted in the last quarter of the twentieth century from the consideration of everyday activity to the effect of explicit physical training in other words: sports.

The question is currently being discussed as to whether and to what extent physical activity counterbalances the elevated risk of mortality of a sedentary lifestyle and whether long periods of sitting should be regarded as an independent risk factor (cf. Bucksch and Schlicht 2013; Rohm Young et al. 2016) ("Sitting as the new smoking"). A recent study shows that daily moderate activity to the extent recommended by the WHO (approx. 30 min of light to moderate activity on at least 5 days per week) is sufficient to compensate for the negative effects of prolonged sitting <8 h. For sitting times >8 h per day, however, significantly higher activity volumes than those mentioned are necessary for compensation (cf. Predel and Nitschmann 2017; Ekelund et al. 2016).

Despite the wealth of evidence outlined above, over 80% of all people over 30 years in Germany have a lack of physical activity (cf. Löllgen and Löllgen 2004).

The majority of working people today spend more than 10 h a day in the context of work, including commuting to and from work.

This fact in combination with the considerations above explains the great health importance of giving employees a more active lifestyle through a modern design of the working environment. Healthy behaviour should be facilitated for everybody not just for those who pay attention anyway ("Make the healthy way the easy way").

Meanwhile, there are various ideas, how this goal can be reached by appropriate measures. There is not one single goal-oriented offer, but a multitude of small steps is required.

7.2 Bicycle-Friendliness

In a recent study, participants who started regular cycling between baseline and follow-up six years later reduced the probability of a narrowing of the coronary arteries by more than 20% (cf. Blond et al. 2016). Professionals who travel to and from work by bicycle are healthier and have fewer days of incapacity to work than commuters who travel by car. Whether the effect is in the range of 30% fewer sick days per year as in a German study (see Kemen 2016) or only 15% as in a Dutch study (see Hendriksen 2009) is secondary for our consideration.

In order to increase the attractiveness for cyclists, certain infrastructural requirements should therefore be met.

7.2.1 Bicycle Parking Facility

Good parking facilities for bicycles should be available in sufficient numbers (current bicycle traffic share plus reserve) and should have suspension points for attaching the bicycles. If the parking facility is located at a point in the public traffic area that is difficult to see, it may make sense to offer heavy stationary locks for permanent use. Camera or video surveillance may also be useful. A roofing of the facility increases the comfort and protects the bikes from the weather.

7.2.2 Changing Rooms and Showers

Depending on the season and the length of the cycle path, cyclists should be able to freshen up and change their clothes. In addition to the possibility of taking a shower, this also includes a practicable offer where the currently unused clothing can be stored, be it in lockers or in changing rooms.

The entire topic is supported by the initiative "Cycle-Friendly Employer" of the EU and national organizations like the ADFC (cf. ADFC 2017).

7.2.3 Staircase and Climbing Stairs

Stair climbing is less perceived by the public as physical activity than cycling. Paffenbarger already showed in the 1970s that men who climb more than 50 steps a day suffer more rarely from heart diseases than men who do not climb stairs and otherwise behave similarly (cf. Paffenbarger et al. 1978, 1986).

In a more recent study, it could be demonstrated that 7 min climbing stairs daily reduces the risk of cardiovascular diseases by more than 50% (cf. Eves et al. 2006).

Accordingly, the question of which measures can effectively motivate people to use stairs is the subject of various studies (cf. Van Calster et al. 2017; Wallmann et al. 2009; Boutelle et al. 2001).

It is important that the stairs are clearly visible, do not have to be searched for and that users are not automatically directed to the lifts. Large, inviting stairwells with wide steps, attractive design and daylight are better used. Illustrations telling a story from floor to floor are also helpful.

If there is an additional indication on the lifts of the health benefits of climbing stairs, the motivation to use stairs is increased.

7.3   Workplace Design and Physical Activity

7.3.1   Dynamic Work Space

In a very recently published paper, it was shown that employees move 32% more in offices with open bench seating compared to closed individual offices. This was proven by a US study carried out in various authorities. One reason for this is presumably the more frequent visit to quiet "meeting points" for discussions with colleagues (Lindberg et al. 2018). Even if the word "dynamic" is usually not used precisely for this aspect of dynamics in offices but more on the fact of moving from one workplace devoted to a special demand to another according to the particular needs at that moment the literal meaning of the word gets into the focus.

7.3.2   Electric Height-Adjustable Desks

In modern offices with the concept of Dynamic Work Space or activity-based working and desk sharing, electrically height-adjustable desks (Fig. 7.1) must be used as standard office furniture because only then is it possible to ergonomically adjust the table height to the biometrics of the user when the user changes at short notice. The usual desks, which should ideally be adapted with tools to ergonomic requirements after consultation with the company doctor or the safety specialist, are in this working environment out of the question.

This also means that all employees in the Dynamic Work Space need to know which height is the right one for them when they are sitting or standing at a desk. Experience has shown that this knowledge is not intuitively accessible to most people.

In addition to this obvious necessity, which also results from the consistent implementation of the directives on VDU-work of several countries in desk sharing, electrically height-adjustable desks also offer the possibility of working alternately in a sitting and standing position, which can increase productivity and lead to less discomfort in the musculoskeletal apparatus of the back and upper extremity (cf. Garrett et al. 2016; Hedge 2004). However, the mere offer of this possibility does not automatically lead to its use (cf. Gilson et al. 2012; Neuhaus et al. 2014). Without instructions as to why and how a new tool should be used, the investment has no broad impact. This has already been experienced by numerous companies who have made a well-meaning switch from standard office desks to electrically height-adjustable ones without offering any training for this measure.

Occasionally, adverse effects of a long-standing position have also been described (cf. Karakolis and Callaghan 2014), but these are not to be expected from reasonable alternate standing and sitting use according to the corresponding instructions.

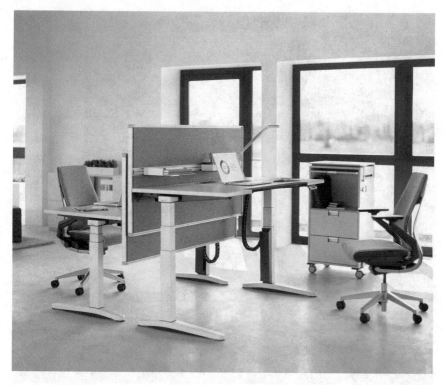

Fig. 7.1 Height-adjustable desks. *Source* Steelcase

7.3.3 Treadmill Desk and Bicycle Desk or Bicycle Seats, Etc.

The treadmill desks that have been available for some time should rather not be regarded as a permanent workplace. The employee stands on a treadmill, which is operated at a low, self-selected speed of 1.5–5 km/h, in front of an electrically height-adjustable standing desk (Fig. 7.2). In the Dynamic Work Space, such an option can be an attractive offer for employees to use intermittently to change position and reduce hours of immobility. It is not intended as a piece of sports equipment, but nevertheless positive effects on the body mass index, on musculoskeletal complaints, even on various metabolic parameters in the blood up to improved cognitive performance have been recorded in studies (cf. MacEwen et al. 2015; Levine and Miller 2007). Individually, however, a reduction in performance in fine motor movements, e.g. lower precision in mouse use, can also be observed (cf. MacEwen et al. 2015).

Similar effects are described with the even less widespread cycling desks (cf. Torbeyns et al. 2014) or the occasional use of a trampoline as a standing opportunity in front of the desk (Fig. 7.3).

Fig. 7.2 Treadmill. *Source* Steelcase

Particularly with treadmill desks and trampolines in offices, their possibly higher accident potential compared with the other variants should also be mentioned, even if relevant experiences have not yet been published.

7.3.4 Movement-Supporting Further Elements

A further possibility for activation, which should be used primarily when furnishing meeting rooms, but also in canteens, are tables that are designed as standing tables or height-adjustable tables, so that shorter meetings of up to approx. 30 min can be held casually while standing, or lunch can also be taken occasionally while standing.

Replacing printers directly at the workplace with central network printers also leads to increased physical activity. These printers are often unpopular, precisely because they lead to having to move away from the worktable. The information that, in addition to other (primary) reasons, health aspects also speak in favour of these solutions can in some cases lead to greater acceptance.

Fig. 7.3   Trampoline. *Source* Bellicon

7.3.5   Employee Information

The offers mentioned above are measures of behavioural prevention, which can unfold their effect only if they are used by employees in the right way. A number of them are either in need of explanation or are clearly better accepted after appropriate communication, so that only then the full behavioural preventive aspect comes into effect. It follows from this that, at the same time as technical assistance is offered, the need for information must be considered immediately.

7.4   Conclusion

There are now effective ideas and options on how to counteract in everyday working life the increasing immobility which evolved in the past decades in the context of the social development in professional and private living contexts with their health-related consequences. For life is movement and movement must not only take place virtually in the mind, but also very literally.

Literature

ADFC. (2017). *Fahrradfreundlicher Arbeitgeber*. https://www.fahrradfreundlicher-arbeitgeber.de/, Zugriff am 10.5.2017.

Blond, K., Jensen, M. K., Rasmussen, M. G., Overvad, K., Tjønneland, A., Østergaard, L., et al. (2016). Prospective study of bicycling and risk of coronary heart disease in Danish men and women. *Circulation, 134*, 1409–1411.

Boutelle, K. N., Jeffery, R. W., Murray, D. M., & Schmitz, M. K. H. (2001). Using signs, artwork, and music to promote stair use in a public building. *American Journal of Public Health, 91*, 2004–2006.

Boyd, J. P., Cullen, C. T., Catanzariti, J., & Oberg, B. B. (Hrsg.). (1950). *The papers of Thomas Jefferson*. Band 8, Princeton: Princeton University Press.

Bucksch, J., & Schlicht, W. (2013). Sitzende Lebensweise als ein gesundheitlich riskantes Verhalten. *Deutsche Zeitschrift für Sportmedizin, 64*, 15–21.

Ekelund, U., Steene-Johannessen, J., Brown, W. J., Fagerland, M. W., Owen, N., Powell, K. E., et al. (2016). Does physical activity attenuate, or even eliminate, the detrimental association of sitting time with mortality? A harmonised meta-analysis of data from more than 1 million men and women. *The Lancet, 388*, 1302–1310.

Eves, F. F., Webb, O. J., & Mutrie, N. (2006). A workplace intervention to promote stair climbing: Greater effects in the overweight. *Obesity, 14*, 2210–2216.

Garrett, G., Benden, M., Mehta, R., Pickens, A., Peres, S. C., & Zhao, H. (2016). Call center productivity over 6 months following a standing desk intervention. *IIE Transactions on Occupational Ergonomics and Human Factors, 4*(2–3), 188–195.

Gilson, N. D., Suppini, A., Ryde, G. C., Brown, H. E., & Brown, W. J. (2012). Does the use of standing ‚hot' desks change sedentary work time in an open plan office? *Preventive Medicine, 54*(1), 65–67.

Hedge, A. (2004). *Effects of an electric height-adjustable worksurface on self-assessed musculoskeletal discomfort and productivity in computer workers*, Technical Report 0904. Cornell University Human Factors and Ergonomics Research Laboratory.

Hendriksen, I. (2009). *Reduced sickness absence in regular commuter cyclists can save employers 27 million euros*. Von: TNO—Knowledge for business. http://www.vcl.li/bilder/518.pdf, February 2009, Zugriff am 10.5.2017.

Karakolis, T., & Callaghan, J. P. (2014). The impact of sit-stand office workstations on worker discomfort and productivity: A review. *Applied Ergonomics, 45*, 799–806.

Kemen, J. (2016). *Mobilität und Gesundheit*. Springer Spektrum.

Levine, J. A., & Miller, J. M. (2007). The energy expenditure of using a „walk-and-work" desk for office workers with obesity. *British Journal of Sports Medicine, 41*, 558–561.

Lindberg, C. M., Srinivasan, K., Gilligan, B., Razjouyan, J., Lee, H., Najafi, B., et al. (2018). Effects of office workstation type on physical activity and stress. *Occupational and Environmental Medicine, 75*, 685–695.

Löllgen, H., & Löllgen, D. (2004). Körperliche Aktivität und Primärprävention. *Deutsche Medizinische Wochenschrift, 129*(19), 1055–1056.

MacEwen, B. T., MacDonald, D. J., & Burr, J. F. (2015). A systematic review of standing and treadmill desks in the workplace. *Preventive Medicine, 70*, 50–58.

Morris, J. N., Heady, J. A., Raffle, P. A., Roberts, C. G., & Parks, J. W. (1953). Coronary heart disease and physical activity of work. *Lancet, 265*, 1053–1057.

Neuhaus, M., Healy, G. N., Dunstan, D. W., Owen, N., & Eakin, E. G. (2014). Workplace sitting and height-adjustable workstations: A randomized controlled trial. *American Journal of Preventive Medicine, 46*(1), 30–40.

Paffenbarger, R. S., Jr., Hyde, R. T., Wing, A. L., & Hsieh, C. C. (1986). Physical activity, all cause mortality, and longevity of college alumni. *New England Journal of Medicine, 314,* 605–613.

Paffenbarger, R. S., Jr., Wing, A. L., & Hyde, R. T. (1978). Physical activity as an index of heart attack risk in college alumni. *American Journal of Epidemiology, 108,* 161–175.

Pagenstert, G. (2017). Bewegt Euch! *Deutsche Zeitschrift für Sportmedizin, 68,* 51–52.

Predel, H.-G., & Nitschmann, S. (2017). Wie beeinflusst körperliche Aktivität die Mortalität? *Internist, 58,* 753–756.

Young, D. R., Hivert, M. F., Alhassan, S., Camhi, S. M., Ferguson, J. F., Katzmarzyk, P. T., et al. (2016). Sedentary behavior and cardiovascular morbidity and mortality: A science advisory from the American Heart Association. *Circulation, 134*(13), e262–e279.

Tipton, C. M. (2008). Susruta of India, an unrecognized contributor to the history of exercise physiology. *Journal of Applied Physiology, 104,* 1553–1556.

Torbeyns, T., Bailey, S., Bos, I., & Meeusen, R. (2014). Active workstations to fight sedentary behaviour. *Sports Medicine, 44*(9), 1261–1273.

Van Calster, L., Van Hoecke, A. S., Octaef, A., & Boen, F. (2017). Does a video displaying a stair climbing model increase stair use in a worksite setting? *Public Health, 149,* 11–20.

Wallmann, B., Mager, S., & Froboese, I. (2009). Treppe statt Rolltreppe. Fördern spezielle Plakate die Treppennutzung? *f.i.t. Forschung – Innovation – Technologie. Das Wissenschaftsmagazin der Deutschen Sporthochschule, 2,* 32–36.

Chapter 8
Outlook Office 4.0

Werner Seiferlein

This last chapter summarizes above all the current technical, but also the organizational state of the office working world, speculates a little about its further development and wants to think about the future challenges with a small look into the glass ball.

Already today, we find a wealth of interconnected technology n, which make our everyday lives more efficient and simpler, but also create new dependencies. Let us just think of the Internet. Some of them predicted in their beginnings only a use by a handful of military men and without which our (working) life today seems almost impossible. This rapid development was only made possible by the technical improvements of, e.g., storage media, fiber optic cables and WLAN and already now gives us the possibility to network at any time and with any place—provided that an appropriate infrastructure is available everywhere. Regions without this technology are formally cut off from the rapid exchange of knowledge and information and can no longer (economically) keep up, which is why, for example, many municipalities in this country are using the fiber optic network in rural areas. The increasing digitalization and globalization and the associated widespread use of the Internet, smartphones, apps, shopping and information platforms will, however, also rapidly influence our consumer behavior, our knowledge acquisition and our ability to make decisions and transfer of knowledge sharing. We are also going to change our forms of social interaction and, of course, our work.

8.1 Focus on the Human Being

The workforce has always been made up of different generations, but never before have their values and needs been so different. The so-called baby boomers born between 1960 and 1969, represent the strongest generation in terms of numbers,

W. Seiferlein (✉)
Technology Innovation Management, Frankfurt/Main, Germany
e-mail: werner.seiferlein@timoffice.de

© Springer Nature Switzerland AG 2020
W. Seiferlein and C. Kohlert (eds.), *The Networked Health-Relevant Factors for Office Buildings*, https://doi.org/10.1007/978-3-030-22022-8_8

are more performance-oriented and attach importance to career opportunities and a high standard of living. The Generation X (1970–1979) finds its way through self-realization and an identification with the job satisfaction at work, while Generation Y (from 1980 to approx. 1995) is characterized by flexibility and a high level of education and also as a meaningful but determined generation is described. For Generation Z, i.e., those born around 1995, a stronger separation of professional and private life as well as even greater flexibility, career and a high income are important. are of secondary importance to them. The generations Y and Z are also called digital natives because digitization has been an everyday occurrence for them since childhood, while earlier generations had to learn how to use computers, the Internet and mobile devices (cf. Scholz 2014; Mangelsdorf 2015).

In offices, there are mostly mixed teams from all generations with their sometimes very different needs for their respective work or workplace. Employers must comply with the flexible working time model. This will also include a new understanding of corporate and employee management as well as work organization. One can use the "Leipzig leadership model" as an orientation, the core of which is the purpose, the sense and reason of a work task or a company. Further dimensions of the model are entrepreneurial spirit in terms of the ability of people, organizations and society to renew themselves, as well as social responsibility. As a fourth dimension, effectiveness translates "responsible and entrepreneurial decisions into goal-oriented strategies, structures and processes to achieve a competitive contribution to the big picture" (HHL o. J.).

Due to globalization, today's world of work is subject to and digitization are subject to constant change. This makes it all the more important in the future to reflect on one's own actions and to communicate and act accordingly. This requires motivated, creative and innovative employees. It is up to employers and managers to create the conditions for this.

8.2 The Change in Office Work

Complex, creative activities, such as the conceptual design of and communication of projects, processes and strategies, will play an even stronger role in office work in the future than in the past. As a result, the creativity of people and teams, their motivation and their well-being are becoming increasingly important.

Up to now, digital media have been used in office work is mainly used to create previously customized digital content from remote team members. and to present them at meetings and workshops, where the cooperation between the of the physically present persons, however, usually takes place on paper (cf. Jurecic et al. 2016). Through the introduction of agile working methods and methods, the need for digital solutions is that support large-format visualization and collaborative editing and development of complex content, but continue to grow. It would be conceivable, for example, to completely digitize the map technology of the programming method (cf.

Kohlert 2017) by means of appropriate software and hardware, which could make the entire process of such sessions more effective and efficient.

In addition, the office space itself will also develop further. Today's diverse technical spectrum enables more and more people to work more flexibly in terms of time and space. In addition to the Corporate Office and the Home Office, so-called coworking spaces can be found more and more, which provide jobs and infrastructure for a limited period of time and in which mainly freelancers are employed, work in mostly larger, open spaces at the same time. Most coworkers are in the creative industry sector or the new media, e.g., as web developer, programmer or graphic designer (see Foertsch 2011b). Coworkers particularly appreciate the flexible working hours and the interaction with other coworkers who do not come from their specialist field (cf. Foertsch 2011a). However, the classic Corporate Office is by no means obsolete, but in the future its spatial and technical equipment will adapt much more strongly to the conditions of modern working methods. In addition to rooms for individual work, there will also be rooms for teamwork and regeneration, all of which will support the development of ideas, creativity, the ability to concentrate and the ability to work in a team or reduction of stress. This does not only concern the furnishing but also the air quality, the noise level or the possibilities of break design in communicative kitchenettes. Especially for large companies with many employees and several, partly overlapping projects there is an enormous potential here, the well-being, the motivation and health of its employees in order to ensure its own economic performance as an attractive employer.

8.3   Future Challenges

The ubiquitous digitalization and the availability of data independent of time and place of various forms on mobile as well as stationary end devices already change almost all processes in the working world today. Several years ago, BASF launched four projects to investigate the effects of digitization on the supply chain, production, research and development (see above 2017). This is not only about the digital creation of documents, but above all about the collection, evaluation and further processing of data by self-learning, increasingly autonomous software systems. In medicine, there is a wide range of tasks for such systems, for example, in cardiac pacemakers or insulin pump n are already in use. In addition, cancer research has developed an adaptive algorithm that can (cf. KYOCERA Germany 2017).

In many places, these technologies have already led to a comprehensive shift of tasks, not only in standardized activities that can be automated more easily, such as the maintenance and servicing of equipment, parts of equipment or plant, but also in more complex and open-ended knowledge work. Here, intelligent algorithms evaluate data according to certain criteria and thus form the basis for decisions, e.g., for creditworthiness checks, through automated trend research, in portfolio management or with the help of parametric planning software in house building.

In the future, more and more tasks will be taken over completely or partially by these software systems. With Basler Insurance, these algorithms already regulate glass damage fully automatically, for example, and employees only check random samples (cf. Engelage 2017). And the Hamburg container port of Altenwerder has a complex control system the running that the envelope with the storage and rail and road traffic on the entire terminal (cf. HHLA n/a).

This further development of networking allows the following game of thought. One of the technologies that is currently overcoming its teething troubles is the so-called smart home, i.e., the control and monitoring of building technology via the smartphone and its networking with the energy service provider, which will soon help to increase energy efficiency and save resources. Connected to wearables, portable sensors that monitor vital signs and which are already used in sport and controlled by an autonomous software system that recognizes the conditions under which we work creatively, concentrated and effectively, our working environment, i.e., the parameters air, light and noise, could be automatically adapted to our state of health and our work task.

The idea of fully automatic optimization of our working environment certainly solves many fears if not rejection. However, since the technical prerequisites are in place for this, in the near future. This has consequences on several levels, not only for office work.

First of all, there are data protection and data security. Responsibility for what data is collected and how it is processed should not be left to self-learning software systems alone. It is the duty of every individual to handle personal information more carefully and responsibly.

The social factor should also not be underestimated. Robotics in particular continues to develop at a rapid pace by optimizing robots in terms of voice, appearance and artificial emotions, which opens up a wide range of completely new applications. One day, for example, the robot might be able to do the physical labor for office activities such as printing scan (cf. Budras 2017). Although this would be more convenient for employees, it would have a counterproductive effect on their energy balance (see Worm 2015, pp. 9 and 17 as well as Chap. 7 Physical activity in the modern working world). Further more, informal meetings, among other things. with colleagues and thus reduces the possibility of communication between employees.

A third point concerns the control of decisions made by artificial intelligence. The algorithms used must not only be comprehensible and logically programmed, but their ethical dimension must also be clarified and defined in advance, because not everything that is legal is also legitimate. The question would be, for example, which ethical rules the algorithm of an autonomously driving car should be controlled according to and whether a random generator decides either to run over a child or possibly to kill the driver, or whether the car is programmed in such a way that it actually performs an electoral action (cf. Jumpertz and Pinkwart 2017).

Fourthly, automation, robotics and artificial intelligence certainly also threaten jobs, as is already the case in industry (cf. above v. 2017). However, with the loss of jobs through the use of robots n and automation, new jobs will be created step by step. Due to the requirements of programming, development, operation, maintenance and machine service of these systems, other work tasks will be created, which still have to be defined and for which, of course, highly qualified employees are required.

8.4 Conclusion

Digitization, automation and development of artificial intelligence are currently taking place without a superordinate structure. Individual industries or disciplines en research and develop more or less deeply and consider further possible applications. These many small activities would have to be brought together in a master plan, which is centrally managed by associations or also by state support, because the networking lives from the integration of these different industries and disciplines.

"The most important thing shall be man John Crone, the father of artificial intelligence and developer of the Watson computer system (cf. Budras 2017"; Watson is able to answer questions of a general nature, link data intelligently, draw conclusions and make forecasts). What is most important, i.e., where the opportunities and challenges of digitization lie, must be discussed on a broad societal basis in order to exploit their opportunities and identify dangers at an early stage. This should lead to the above-mentioned master plan, which both launches an education offensive for new workers and involves all concerned, takes away fears and develops positive visions to answer the pressing questions of the future.

Literature

Budras, C. (2017). Rock star of IT. In *Frankfurter Allgemeine Zeitung*, business section, 9.6.2017, p. 26.

Engelage, H. (2017). *Where the robot already regulates damage*. http://www.gdv.de/2017/03/where-the-robot-is-regulated-in-a-kind-shared. Accessed 8.10.2017.

Foertsch, C. (2011a). *What coworkers want*. http://www.deskmag.com/de/welche-coworking-spaces-coworker-wollen-165. Accessed on January 12, 2018.

Foertsch, C. (2011b). *The coworker profile*. http://www.deskmag.com/de/die-coworker-global-coworking-survey-168. Accessed on January 12, 2018.

HHL Leipzig Graduate School of Management. (no year). *The Leipzig management model*. https://www.hhl.de/de/hhl/leipziger-fuehrungsmodell/#1. Accessed on January 10, 2018.

HHLA Hamburger Hafen und Logistik AG. (no year). *As if by magic*. https://hhla.de/de/container/altenwerder-cta/so-funktioniert-cta.html. Accessed on January 10, 2018.

Jumpertz, S., & Pinkwart, A. (2017). *The core is the purpose*. https://www.managerseminare.de/ms_News/Guide-model-for-the-future-in-the-core,252264. Accessed on January 10, 2018.

Jurecic, M., Rief, S., & Schullerus, M. (2016). *Digital work—Motifs and effects of low-paper working methods*. Stuttgart: Fraunhofer.

Kohlert, C. (2017). Creation of another user description. In W. Seiferlein & R. Woyczyk (Eds.), *Project success—the networked factors of investment projects* (pp. 55–78). Stuttgart: Fraunhofer.

KYOCERA Germany. (2017). *The Japanese technology group KYOCERA and the University of Tsukuba are now developing an AI-based image recognition system for skin diseases*. http://www.kyocera.de/index/news/previous_news/news_archive_detail. L2NvcnBvcmF0ZS9uZXdzLzLzIwMTcvRGVyX2phcGFuaXNjaGVfVGVjaG5vbG9naWVWrb G256ZXJuX0tZT0NFUkFfZW50d2lja2VsdF9qZXR6dF9taXRfZGVyX1VuaXZlcnNpdGFldF9U c3VrdWJhX2XVuaXZlcnNpdGF0dWJhX2VrdWXX2html. Accessed on January 5, 2018.
Mangelsdorf, M. (2015). *From baby boomers to Generation Z: The right way to deal with different generations in the company*. Offenbach: Gabal.
o. V. (2017). Digitalization threatens jobs in the chemical industry. In *Frankfurter Allgemeine Zeitung*, business section, 27.9.2017.
Scholz, C. (2014). *Generation Z: How it ticks, what it changes and why it infects us all*. Weinheim: Wiley-VCH.
Worm, N. (2015). *Happy and slim*. Lunen: systemed.

Chapter 9
Summary and Outlook

Werner Seiferlein and Christine Kohlert

9.1 Future Office Forms

Employees, like most people, want to be part of a community, to feel belonging and comfortable, and for this they need a harmonious working environment that supports them in their work in the best possible and holistic way. Everyone is talking about well-being and the right support at work, and this is the prerequisite for a new working environment in which employees are motivated, can perform well, make the right decisions quickly and well, and in which communication and exchange within the team work well. Employers have a duty to take care of their employees and to provide them with the best possible support in fulfilling their tasks. They play the role of a "caretaker" who also cares about their health, both physically and mentally and supports employees in maintaining or, if necessary, improving their health.

The various topics described by experts in the previous chapters are decisive for the success of a healthy working environment and how it is perceived.

Of course, there are always illnesses in offices and they have different causes. This book was written to find out what causes diseases and how to remedy some, alleviate others or avoid others altogether. It is to be understood as a guideline for action and as a checklist which should be used as early as the planning stage.

The goal of the development of office buildings is based on different intentions. Just a few years ago, the transition from cell to open space was primarily about the economic aspects. This step enabled space savings of 30–60%. The introduction of a shared desk will further reduce this cost block. These values are based on the

W. Seiferlein (✉)
Technology Innovation Management, Frankfurt/Main, Germany
e-mail: werner.seiferlein@timoffice.de

C. Kohlert
RBSGROUP Part of Drees & Summer, Munich, Germany
e-mail: christine.kohlert@rbsgroup.eu

© Springer Nature Switzerland AG 2020
W. Seiferlein and C. Kohlert (eds.), *The Networked Health-Relevant Factors for Office Buildings*, https://doi.org/10.1007/978-3-030-22022-8_9

statistical assumption that due to holidays, training, illness, etc. the workforce is not present for about 10–20% of their working time.

Today, other aspects are coming to the fore. It is much more about networking, communication, exchange and the right teamwork as well as quick decisions and having the knowledge available when you need it.

Based on the data available today, "open space" is more or less accepted. In the "shared desk" office form, the opinions of the experts diverge, and the clean desk policy is also rejected by the overwhelming majority of employees. The criticism concerns above all the daily clearing away of all documents and private things in order to be able to allocate new jobs the next day (Bös 2017). Also the individualization of one's own work table or working environment plays an important role for people. Personal items such as family photos or plants but also equipment and utensils for daily use should be left at the table.

9.2 Future Health Planning

In addition to the rational aspect of reducing the floor space required per employee, the home office also has a psychological aspect. As a rule, it ensures that things can be processed in peace and quiet that the retreat zones of the office concept often cannot achieve in this way. In addition, an "imaginary" distance to the company workplace is created. This consideration is also supported in the context of mission investigations. In general, business trips with an average of 25–50% of working days have a positive effect on well-being, motivation and performance.

In Sect. 10.1 are the checklists for the individual chapters that refer to the health-relevant factors that should be taken into account when planning an office building.

The factors are partly interlinked. The influence and interdependencies of these factors on each other were determined in the book and can thus be incorporated into the early planning phase in the future. Examples of networking include colours and light, perceptions and agreements (e.g. noise), agreements and colours, needs-based office buildings and perceptions, robots and office furnishings (see Table 9.1).

The office environment could be shown throughout the book as a strong factor for the "well-being of the employees" (furnishing, acoustics, retreat possibilities, break areas, colour concepts, fresh air, etc.). Chap. 6 "Medicine ische Aspekte" is a very important factor and closely interlinked with all the chapters and factors dealt with here in the book.

Odour assessment methods are state of the art in the industry. The spectrum of applications ranges from the detection of harmful vapours (cf. DIN ISO 16000-28 2012) to the conscious scenting of products to produce a product-specific, acceptable odour sensation. The addition of "more or less" clearly perceptible fragrance. However, the use of stand-alone devices in the air does not necessarily lead to the desired results.

Table 9.1 Networking the health-relevant factors of an office building

	Chapters	Arrangement	Perception	Colours	Office furnishings	Building technology	Medicine	Movement	Office 4.0
1	Arrangement		X				X		
2	Perception	X		X	X	X	X		X
3	Colours		X				X		
4	Office furnishings		X				X	X	X
5	Building technology		X				X		
6	Medicine	X	X	X	X	X		X	X
7	Motion				X		X		X
8	Office 4.0		X			X	X		

9.2.1 Health and Well-Being

There is a risk that many health-relevant factors may fail to achieve the desired goal of well-being if the relevant processes, equipment and application are not adhered to (see Sect. 10.3 causes and triggers of diseases).

Gender-specific differences seem less noticeable than expected; the results of the questionnaire do not reveal any clear gender-specific differences. In addition, managers in particular seem to find interruptions in work very burdensome, which should be taken into account especially in the Open Space office (Gabrysch 2017).

In principle, health-related issues should be taken into account both in the planning phase and in the operating phase. Support from management and honest and open communication are in the foreground. This already starts with the planning and ends with the implementation of the process, and the selection of equipment and agreements that are to be implemented within the framework of change management must be implemented and applied accordingly. Potential tensions and thus stress are avoided by "leading by example" agreements, especially by superiors.

Especially, the $4 \times$ Ls (light, air, noise, body) are always on the list of defects of the employees. The medical term "RSI" means Repetitive Strain Injury: Injury caused by repetitive stress. This causes complaints in the neck and shoulder area as well as arm and hand complaints up to specific diseases such as tendosynovitis or carpal tunnel syndrome.

In Fig. 2.2, the corresponding values that contribute to comfort through the air were listed (air velocity, humidity and air temperature).

Aggressive colour landscapes, for example, claim our perception to be more than harmonious, are tiring in the long run and can cause headaches. In order to increase the well-being of the employees, it is desirable to create a harmonious colour concept in which all components, such as carpets, walls, furniture, etc., are coordinated with each other.

The overall aim is to create a dynamic office environment that encourages employees to work in different postures and move around the office. For example, well-designed staircases motivate employees to use them and thus integrate movement into their daily work.

Ergonomics forms the methodical basis in all areas of the furnishing of working environments because long monotonous sitting leads to reduced blood circulation which supplies the muscles with less oxygen. After some time, this leads to muscle hardening, tension and pain. If you are in pain, you tend to be more gentle, the one-sided strain of which can ultimately lead to muscle hardening again and ultimately to muscle atrophy. The musculature loads thereby the spinal column and beyond that intervertebral discs, tendons and ligaments. Particular attention must be paid here to the correct use of the screen, keyboard and mouse as well as the table and chair, and to training the employees to adjust the backrest, seat, armrest, etc. correctly.

Healthy materials must be used in the construction and design of buildings (vapours, hygiene, allergies, etc.). For example, different wood qualities and quantities can cause a significant decrease in diastolic blood pressure and a significant increase in pulse.

Dry air can be humidified by injecting water, i.e. humidifying the air. The situation can be improved. A good alternative is green walls or plant beds, which improve the quality of the air in a natural way.

9.2.2 Cradle to Cradle

Production sites whose "wastewater" has drinking water quality, and clothing that is compostable or becomes food for plants and animals. Carpets, for example, can be activated for reuse before use for refurbishment, or the material used can be reused after use or composted without harmful residues.

Cradle to Cradle follows the principle of thinking in complete product cycles right from the start. The aim is to avoid garbage and to recycle goods at the end of their useful life and to reuse all components as best as possible. It is about thinking in continuous cycles, understanding products and materials as nutrients and leaving a positive footprint of one's own as a human being.

9.2.3 Optimization of Operating and Maintenance Practices

Taking the operation and maintenance of a building into account during the preliminary design phase of a plant leads to improved working environments, higher productivity, reduced energy and resource costs and the avoidance of system failures.

Depending on the type of settlement, whether investor or self-financing, there are different types of execution. The total cost of ownership includes the total costs incurred, for example, in the construction of a new administration building, i.e. both the investment costs and the operating costs.

All parties involved in the project must be involved via the target situation. Contractors and maintenance personnel must participate in the design phases to ensure optimum operation and maintenance of the building.

Designers can specify materials and systems that simplify and reduce maintenance requirements. It is desirable to use less energy and chemicals; this is cheaper and reduces life cycle costs. Sustainability initiatives are desirable, including the reduction of energy and water use and waste generation.

9.3 Future Office Organization

What's the next step? PopTech Curator Andrew Zoli answers the question of where people work best with: "not in the office" (Hofmeister 2017, p. 3; here too, the question of the need for offices is asked. Will the employees only work mobile—no matter where they are?).

The promise of flexibility and personal time management attracts employees to the job but that alone is not enough to keep them there. Young creative employees need commitment, interesting and surprising spaces as well as experiences and an inspiring environment. In order to emotionally bind employees to the company, it is of great importance to make one's own brand values tangible and to transport them internally through an authentic design of the working world. Work spaces that convey basic human needs such as security, recognition and self-fulfilment are emotionally binding and enable free thinking and the necessary attention.

The good mood makes us more compatible, healthier and encourages us to think more creatively. An optimistic and authentic environment that gives employees choices and control and supports movement and interaction, also contributes to this. In addition to these spatial aspects, it is also about meaningful activities and mutual attentiveness and about providing employees with places for exchange and cooperation as well as for peace and concentration.

The office becomes a marketplace for the exchange of knowledge and teamwork. Concentrated work for those who want it takes place there as well, others relocate their activities to various places—third places, such as means of transport (public, self-propelled cars), coworking spaces or even the home office. This can lead to a reduction in significant area proportions. If one assumes, for example, that the employee spends two or three of five days a week in other places, an office building, for example, instead of 1000 workplaces, only has to create about half as many jobs, i.e. only 500. These sharing ratios must be evaluated in advance of planning and determined on the basis of working methods and processes.

In addition, there is a rapid change in technical possibilities. With the introduction of smartphones, tablets, WLAN, etc., the end of the office has already been predicted (ibid., p. 5).

The cons will be:

- conflicts in the management of employees
- compensation of the ergonomic equipment in the home office
- personal communication suffers
- presence must be planned more strongly

The pros will be:

- cost savings
- possible reduction of area
- reduction of the journey to the office building

  - time
  - financial resources

- stress-free everyday life

  - self-determination of everyday life

- fewer interpersonal quarrels (mobbing, quarrel etc.)
- self-determined workplace (creative, relaxing, stimulating) in nature, library, café, etc.

When comparing the pros and cons, there is a conflict. On the one hand, you have the freedom to work where you want in the future, on the other hand the boundaries between work and leisure are blurred. The workplace is no longer the office but also the home, the park, the café and much more.

The campus of the future is characterized by openness. For example, on the ground floor, there is a café, kiosk, parcel service, concierge and much more. This transforms the company-oriented area into a social meeting place for employees and their friends and relatives (example: BMW headquarters in Munich).

Good rooms are always a contribution to the company value. Like in a good restaurant the overall harmony counts, from the welcome to the farewell of the guest. The entire office with its environment and organization makes a decisive contribution to a harmonious working environment that leads to well-being, satisfaction and health of the employees and thus ensures the success of a company.

Literature

Bös, N. (2017). No more tidying up. In *Frankfurter Allgemeine Zeitung*, 11./12.3.2017, p. C1.

Gabrysch, V. (2017). *Job satisfaction in an open space office. An empirical study in a large pharmaceutical company* (Unpublished master thesis). Pädagogische Hochschule Schwäbisch Gmünd.

Hofmeister, S. (2017). *The office of the future is developing into something completely new*. Event in the Bavarian National Museum, Mars-Venus-Saal, 9.3.2017.

Chapter 10
Checklists, Regulations and Suggestions

Werner Seiferlein and Christine Kohlert

10.1 Checklist for Networked Health-Relevant Factors for Office Buildings

10.1.1 Mutual Agreement

- Rules for general behavior are drawn up and applied (see Sect. 10.4, Dealing with colleagues (rules); Clariant 2015), e.g., for telephoning and using retreat rooms, but also for dealing with each other?
- Development of a code of conduct shortly before moving into the new working environment.
- Eye relaxation exercises are performed regularly (Heidenberger 2017).
- When making calls, the handset is not clamped between the head and shoulder (ibid.).
- When working with a PC, the feet stand firmly on the floor. Legs are not crossed (this disturbs blood circulation) (ibid.).
- Prescribed visual aids are used (ibid.).
- The sitting position is changed as often as possible (ibid.).
- Discussion of elements of health-promoting office design.
- Survey of the requirements for the future office environment on the part of the users.
- Answering the urgency question: Why must the familiar and trusted be abandoned if everything works well?

W. Seiferlein (✉)
Technology Innovation Management, Frankfurt/Main, Germany
e-mail: werner.seiferlein@timoffice.de

C. Kohlert
RBSGROUP Part of Drees & Summer, Munich, Germany
e-mail: christine.kohlert@rbsgroup.eu

© Springer Nature Switzerland AG 2020
W. Seiferlein and C. Kohlert (eds.), *The Networked Health-Relevant Factors for Office Buildings*, https://doi.org/10.1007/978-3-030-22022-8_10

- Development of a powerful vision with the management, which leads all steps in the project.
- Form a strong leadership coalition that is aware of its role model function and consistently guides its employees through change.
- Develop a professional change management program that takes into account the specifics of a company and its workforce.
- Managers receive differentiated development offers in order to convey the vision of the company management to the workforce.
- Comprehensive project communication provides employees with security and integrates current information requirements.
- Employee representatives (change agents) and recognized personalities are used as multipliers in the project and involved in sub-projects to design the future office environment.
- Various change management measures reach employees on a broad basis.
- Pilot areas allow the staff to experience the new office concept at an early stage in the project, while test use provides insights before implementation throughout the company.
- Create profits for the workforce throughout the change process that support the vision of management.
- Anchoring the (desired) corporate culture throughout the entire change process and in the design of the new office environment.
- Does my office environment have sufficient space for both communication and concentration?
- In addition to acoustics, climate and light, was the unions accompanied properly in the change process? it important to accompany the MA in change?
- Consider integration of different age groups (young parents—but also care of the elderly).
- Employeesto allow free time management.
- Take care of health care (offer of e-bikes for example).
- Promoting movement in the office (attractive stairwells, only one waste-basket, one printer at a central location, etc.).
- Plan a kitchenette as a multifunctional room.
- Transport the brand values of the company to the inside.
- Develop a coherent culture and vision for the company.
- Use sustainable materials.
- Integrate green (plants).
- Leadership with trust–goal achievement not presence culture.
- Offer changing job opportunities: standing, sitting, resting, lying….
- Employee surveys on satisfaction with the working environment offer adjustment ideas for increasing acceptance.

Worklife Balance

- There are sports rooms for fitness, yoga, or back training available.
- There are childcare facilities.

- Possibility for a home office is given.
- Planning of well-being and satisfaction:
  - Principle 1: There are no guarantees, but spaces are needed for different activities such as concentration and communication. These rooms support creativity and learning and allow employees to make their own choices.
  - Principle 2: Comfort is the key (Fig. 1.9).
  - Principle 3: Space can release good behavior and make us aware of who and what is present. Naturally exposed places for communication and exchange with a view of the countryside promote this attitude.
  - Principle 4: Flexibility and variability are an absolute necessity in order to be able to react quickly and efficiently to changing conditions.
  - Principle 5: Space in connection with nature is worth striving for, i.e., green, as best it can be, to be brought to the inside and to make views possible.
  - Principle 6: A room is only as good as those that lead in it. Ultimately, management determines how many feel-good factors are implemented and how they are used—mutual trust is the most important thing. It is a matter of perfectly translating the relationships that are important for the achievement of corporate goals into space. Good rooms are always a contribution to the company value.

10.1.2   Individual Perception

- Air conditioning/heating can be adjusted individually.
- Ensuring that heating of rooms by solar radiation is avoided at all times, including outside working hours, by lowering blinds or using other sun protection devices.
- Effective control of ventilation equipment, e.g., window opening for night cooling.
- Ventilation in the early morning hours.
- Reduction of internal thermal loads, e.g., electronic devices only to be operated when required.
- An air humidifier (e.g., indoor fountain) is used during the heat period (Heidenberger 2017).
- Plants are available to provide a better indoor climate in the office (ibid.).
- The office can be ventilated regularly (ibid.).
- Different scents are available and can be sprayed into the air.

Noise

- There are closed rooms close to the workplace (acoustically completely separated), which serve as a retreat for concentrated individual work, bilateral meetings, and telephone calls.
- There are lounges that are acoustically and optically shielded from the workstations and serve as a place for concentrated work and regeneration.

- Think tanks and meeting rooms are soundproofed so that you can concentrate on your work (Martin 2006).
- The entrance to meeting rooms is not in the office to avoid constantly changing audiences and through traffic (ibid.).
- Doors are provided and can be opened and closed to ensure concentrated work (ibid.).
- The office has a good noise insulation, so that no noise from outside disturbs the work (ibid.).
- Partition walls are available for noise insulation (ibid.).
- Devices such as printers, copiers, scanners, fax machines, blinds and air conditioning systems are noise-friendly (ibid.).
- Selected music is played in the toilets, staircase, and foyer.
- Headsets are made available to the employees and they are instructed in the correct use and adjustment by the specialist personnel.

Light

- Blinds/curtains are present and function perfectly (ibid.).
- Light can be switched individually (ibid.).
- Sun protection is available against strong solar radiation (Heidenberger 2017).
- The lighting on the desk can be adjusted in steps (ibid.).
- The light sources are located on the side of the monitor or directly above it on the ceiling (ibid.).
- The entire office space is uniformly illuminated (ibid.).
- There is sufficient natural light in the office (ibid.).
- Workplaces must be located close to windows in order to provide a high proportion of daylight (daylight ratio of >4% at all workplaces).
- The lighting should consist of at least two components.
- The lighting should be adjustable by the user.
- The lighting must have both direct and indirect lighting.
- For office workplaces, at least 500 lx must be observed.
- Lighting with light colors adapted to the time of day should be aimed for (human dynamic light).
- Glare or reflections caused by daylight incidence shall be provided with adjustable light protection devices.
- Light is part of a color concept.

Body

- There are spatially and acoustically separated beverage stations.
- There are cafés with a selection of healthy foods (Steelcase 2008).
- There are enough toilets available (share rate).
- What drinks will be available? Water? Coffee? Tea?
- Is there any fruit? Apples? Seasonal fruit?
- How is the food supply for the employees structured? Outside the home? Internal?

- In the case of a canteen: Is healthy cuisine offered with healthy ingredients?

10.1.3   Colors in General and in Particular

- Colors in the light make the objective world visible to us. Matching the colors in the room to the lighting concept is, therefore, an important prerequisite for creating pleasant working environments.
- Monochrome (Monochrome: "Light emission in a very narrow frequency or wavelength range, a perception reduced to only one color receptor, tone-in-tone painting, painting in only one color direction, generally an image or photograph showing only grey scales or gradations of a single color, see black and white," https:// de.wikipedia.org/wiki/Monochrom) as well as uniformly bright environments are unnatural, room concepts with brighter and darker, brighter and shaded, warm and cool as well as colorful and uncolored zones are natural. They have a physiologically relieving effect.
- You see bright things first. The places and areas that are to attract the most attention are to be designed brightly.
- Monochrome camouflages and connects, polychrome shows and differentiate. Both extremes are permanently stressful.

10.1.4   Adequate Furniture

Design of the offices

- Provide spacious and well-equipped rooms that allow both mobile and fixed-site staff to work individually or in teams (ibid.).
- Create spaces that can be individually personalized and adapted and do not force employees into rigid workplace standards (ibid.).
- In the offices, there is enough storage space available (Martin 2006).
- Work with objects that are of high quality, have an appealing or motivating design, that you like to have in front of your eyes and take at hand. This also promotes the urge to order.
- Ensure transparency so that people can see and be seen—this creates a relationship of trust (Steelcase 2008).
- Walls and other vertical surfaces are used to visualize thought processes and progress (ibid.).
- Offer rooms that have a calming effect through materials, surfaces, colors, light and views (ibid.).
- Design areas that allow your employees to decide how many and what stimuli they want to perceive through their senses and withdraw if they feel exposed to too many stimuli (ibid.).

- The workstation is aligned so that there is a window on the right or left side (Heidenberger 2017).
- Decentralized or central print service zones are available to protect workplaces from emissions such as noise, particulate matter and ozone.
- There is a central archive for the office level as a supplement to the personal storage space at the workplace.
- A group storage is located near the associated work centers.
- There are central cloakrooms with lockers as a supplement to the cloakroom hooks at the workplace and the visitor cloakroom.
- There is a centrally located open waiting area which can be used as a reception area, a waiting area for guests, presentations, or as a communication area.
- A lockable personally assigned subject is provided for each employee.
- For reasons of noise protection, photocopiers and printers should preferably be housed in separate photocopying rooms.
- Printing/copying should be possible on every device in the building with its own batch.

Plants—which are suitable "living" plants?[1]

- Sansevieria (Fig. 10.1).
- Gum (Fig. 10.2), Klivie (Fig. 10.3).
- Yucca palm (Fig. 10.4).
- Ficus Benjamin (Fig. 10.5).
- Green lily (Fig. 10.6).
- Room linden (Fig. 10.7).
- Bonsai (Fig. 10.8).
- Efeutute (Fig. 10.9).

Evergreen walls without earth

- Dragon tree (planting wall) (Fig. 10.10).

Avoid dry room air (see Weber 2017).

- The lower limit of the comfort zone is between 40 and 45% humidity.
- The risk of mold growth increases from 60% onwards.
- Weekly cleaning counteracts the risk of bacterial contamination.

Ergonomics

Screen

- With the head tilted, the highest line on the monitor can be read. An angle of approx. 35° below the horizontal visual axis is optimal (Heidenberger 2017).
- The screen is placed in such a way that it can be viewed directly without having to rotate the upper body (ibid.).

---

[1] The photographs of the plants and the plant wall were taken by Christine Kohlert.

Fig. 10.1 Bow hemp

- The screen resolution is set so that even small fonts can be read without problems (ibid.).
- A black font is used on a white background.
- The brightness and contrast of the image are adapted to individual needs (ibid.).
- A document holder is used for typing documents (ibid.).

Keyboard and Mouse

- Provide each employee with his or her own keyboard (Reimersdahl no. J.).
- When operating the keyboard, the arms on the desk surface form a right angle to the upper body (Heidenberger 2017).
- The maximum height of the keyboard at its highest point is 3 cm. If no, a base is used for the arms (ibid.).
- In front of the keyboard, there is 5–10 cm space for the palm rest (ibid.).
- The keyboard is easy to operate (pressure sensitivity of the keys).
- Ensure that keyboards, PCs, and printers are cleaned regularly (with compressed air) (Reimersdahl no. J.).
- A palm rest or ergonomic mouse is used to operate the mouse (Heidenberger 2017).
- When operating the mouse, the forearm can be placed comfortably on the desk (ibid.).

Fig. 10.2 Gum

Desk

- The forearms can be placed relaxed on the desk without having to pull the shoulders or bend the back (ibid.).
- Sufficient free working space is available (ibid.).
- The desk is fixed and free of sharp edges or dangerous corners (ibid.).
- There is sufficient distance between the knees and the table top (ibid.).
- The desk is about 75 cm high (ibid.).
- The telephone is within easy reach (ibid.).

Desk Chair

- Shank and thigh form an angle of about 90° when sitting (ibid.).
- A footrest is used to improve the leg posture (ibid.).
- With a hard floor covering (e.g. laminate): The armchair has soft castors (ibid.).
- For a soft floor covering (e.g. carpet): The armchair has hard castors (ibid.).
- The armchair is rotatable (ibid.).
- The height of the armchair is adjustable (ibid.).
- The backrest adapts flexibly to movements (ibid.).
- The office chair has a suspension that absorbs sitting down (ibid.).
- The office chair has adjustable armrests (ibid.).

Fig. 10.3 Glivie

Office Furniture

- Each storage compartment can be easily reached (ibid.).
- A safe stool or "elephant foot" is used to achieve higher storage possibilities (Heidenberger 2017).
- Other office furniture is free of defects such as jamming drawers or loose shelves (ibid.).
- Provide easily adaptable furniture to support different sizes, needs, and preferences and allow movement in the workplace (Steelcase 2008).
- Classic workstation areas: Stand-sit table, electrically height-adjustable with height display to simplify adjustment; screens to inhibit sound propagation, absorb sound, and protect against visual interference; ergonomic office swivel chair with intuitive adjustment options; docking station; keyboard; mouse; monitor holder; (2) screen(s).
- Short-term workstations (touchdown workstations): electrified bench solution for up to a maximum of eight people with acoustically effective screens and, if necessary, wireless charging; ergonomic office swivel chairs with intuitive adjustment options.
- Rooms for isolated and confidential or concentrated activities, individually or in groups: Room-in-room systems (think tanks) with individual lighting and air-

Fig. 10.4 Yucca Palm

Fig. 10.5 Ficus

conditioning control, furniture that is appropriate for the majority of uses (here, several options are often made available with different equipment variants).

- Workshop rooms: flexible tables, foldable and/or on castors, and plug and play electrification; flexible chairs (low weight, on castors); smartboards or whiteboards; storage space for workshop materials.
- Creative or project spaces: Table in standing height; benches or standing aids; smartboard/monitor with wireless connectivity; whiteboard.
- Cafeteria/Bistro: varied, attractive seating at gastronomic level; different table variants (different heights and respective number of seats).

Fig. 10.6 Green lily

Fig. 10.7 Room linden

Equipment for Better Communication

- Provide video conferencing configurations that allow non-physically present participants to track and see content and understand everyone equally (Steelcase 2008).

10.1.5 Required Building Services Engineering

- Health in real estate is measurable, plannable, feasible, and affordable. This has been successfully tested in many hundreds of projects with several thousand units.
- The most important criterion is the pollution of indoor air with pollutants, for which there are official criteria.
- The term "healthy" should be treated with caution for scientific and liability reasons.
- "Eco" or "organic" is not automatically healthy.
- Emphasis should be placed on an exact definition of terms (e.g., sustainability).
- As a consequence of the ECJ ruling on the German approval of building materials, the tested (health) quality of building products has a new significance.

Fig. 10.8 Bonsai

- For those who already build and renovate well in terms of quality, it is only a short way to the tested and healthier buildings.
- With model projects, a healthier building can be implemented well in the company. TÜV Rheinland is a valuable and experienced partner for the evaluation and certification of building materials, the testing of buildings, and the training and certification of employees.
- The Sentinel House concept comprises several stages of qualification, building material control, construction supervision, and final measurement. To this end, a comprehensive advisory, training, and further education programme were developed for all actors in the construction sector; www.sentinel-haus.eu.
- Technologies are ready to display information in real time (Steelcase 2008).
- For cabling over traffic routes and movement areas: Are cable bridges provided?

Healthy material and maintenance

- Contractual target agreement for defined pollutants between client and contractors.
- Qualified specialist planners for tenders, conditions, and construction systems.
- Qualified skilled workers know building materials and processing conditions.
- Selection of building materials according to health criteria with criteria including advice from the specialist trade.

Fig. 10.9 Efeutute

Fig. 10.10 Dragon tree

- Use of quality management and quality assurance.
- Independent investigation and analysis according to qualification (Bachmann and Lange 2013).
- Barrier-free design, design of workplaces for severely disabled people (quality of floors. no tripping hazards, anti-slip, even and easy to clean).

Considering all health risks, for example:

- Air pollution
- Water pollution
- Heat island effect
- Noise exposures
- Electromagnetic pollution
- Risks associated with related activities
- Security of individuals and their property
- Danger of epidemics
- A plan to take these man-made risks into account and an assessment of the extent of these risks should be prepared
- Analysis of natural risks and determination of the extent of these risks
- Given the increased frequency and severity of natural meteorological and geological events, natural risks should be analyzed for each proposed site
- fires
- Tornados–Windstorms
- Heat waves
- Extremely cold
- Floods–torrential rains
- Earthquakes–landslides–volcanic activity
- Diseases related to the natural environment
- A comprehensive prevention plan should be drawn up, together with an assessment of the risks and the extent of those risks.

10.1.6 Medical Aspects

Hygiene

- Observe hygiene, cleaning the workplace, and hygiene dispenser.
- For hygienic reasons, the keyboard is not shared with other colleagues.
- Showers are available.
- Provide disinfectant wipes so that you can wipe off the telephone yourself (Reimerdahl no. J.).
- Make sure that desks are cleaned thoroughly on a regular basis and that rubbish bins are emptied (ibid.).
- Regular hand washing is the order of the day and requires disposable towels (ibid.).
- Prohibit your employees from eating at their desks. Instead, create a central meeting point in the office where employees can drink coffee, eat something and exchange ideas or news with colleagues in between (ibid.).
- General office rules should also cover the area of hygiene for the benefit of all. The following guidelines are examples of how they could be defined:

- Once a week, clean the keyboard, telephone receiver, and computer mouse with a fat-soluble detergent (in many companies these things are also personally assigned to an employee).
- Air the work areas several times a day at fixed times to reduce the number of viruses and bacteria in the room by exchanging air.
- Wipe the refrigerator regularly with hot water with cleaning additive.
- Dispose of expired food.
- Wash hands regularly and provide disinfectants in washrooms.

Allergies

- Germ inhibiting coated surfaces [cf. Chap. 4 (avoid adequate furniture, dry indoor air)].

10.1.7 Movement in Regard to Workplace

- In addition to the statutory minimum widths, the main traffic routes are also designed according to qualitative aspects such as multifunctionality.
- Regular breaks are also used for more movements (e.g. walking through the office) (Heidenberger 2017).
- Are sufficient numbers and quality of bicycle parking facilities planned?
- Are changing facilities, showers, and lockers provided?
- Are the stairwells good and immediately visible?
- Are the stairwells attractively designed, airy and inviting?
- Are the office furniture state-of-the-art, including electrically height-adjustable desks?
- Has it been checked whether treadmill desks or bicycle desks can be used?
- Are some meeting rooms/partitions of the canteen equipped with bar tables?
- Are the employees informed in detail about the use of the hardware?
- Do the employees know of any additional health-relevant behavioral options beyond the mere use of the devices?
- Activity-Based Working: Offer mats, motion chairs, possibly rings for "hanging."

Additional behavioral tips:

- Park your car further away or get off two stops earlier.
- Stay on public transport (of course with your hand on the handle).
- Ride your bike (especially in good weather) to work, if the road is too far, consider combining train and bike.
- Use the stairs instead of the elevator.
- Change sitting and standing positions more often—the next position is the best one.
- Use the dynamic sitting setting of the desk chair.

- Find out how to adjust the desk chair ergonomically for yourself, the same applies to the height-adjustable desk.
- Take regular exercise breaks.
- Clarify not everything by phone or e-mail, visit your colleague sometimesclarify, calmly also times personally in the office of the colleague visit and thereby stretch the legs stretch.
- Make calls standing up or walking around.
- Unleash your creativity while standing or walking around.
- Hold meetings standing or walking.
- Do exercises at your desk that help against tension.
- Use Desktop-Reminder for more movement.
- Use the breaks (especially the lunch break) also as exercise breaks. Out of the office into the fresh air—not only good for the musculoskeletal system, but also for the head.
- Motivate employees and colleagues to implement the above behavioral tips.

10.1.8   Office 4.0

Perspectives on the open space office (Weber 2017)

- Visionary: It is about the future of work and business. Has his own visions. Think big: perspective. Silicon Valley,
- Modernizer: Knows that the world is changing and that one's own company needs to change. Has ideas, but no visions, drives less, reacts more. Goal: Change.
- Pragmatist: Realizes what others have decided. He knows that visions and spaces are not the same and people are complicated. Experience: One always grumble.
- Doubter: Finds that something should change, but not necessarily so. Mistrusts the motifs of the makers and the pretty pictures. The change should be different: more honesty, more participation. Assumption: It is all about the costs.
- Opponent: The fact that it is only about money is no question for the opponent—it is exactly the same, and at the expense of his health and performance. Charge: I cannot work like this.

Objectives for change (Kratzer 2017)

- Economic motives
- Communication
- Accelerate information flows
- Shorten coordination processes
- Better distribution of knowledge
- Informality Surplus
- Attractiveness
- Flexibility.

Factors for change (Bachmann and Lange 2013)

- Attitude: credibility, consistency
- Participation: Co-determination, employees, managers
- Competence: Comparisons, consulting, analysis
- Control: project-based, interdisciplinary, piloting.

10.2 The Most Common Regulations (Professional Associations, Workplace Guidelines, Etc.), Rules and Information (See Chap. 4)

- Occupational Health and Safety Act (ArbSchG)
- Workplace Ordinance (ArbStättV)

  - ASR A1.2 Dimensions of rooms, airspace
  - ASR A1.3 Safety and health protection marking
  - ASR A1.5 Floors, walls, ceilings, roofs
  - ASR A1.6 Windows, skylights
  - ASR A1.7 Doors, gates
  - ASR A1.8 Traffic routes
  - ASR A2.2 Measures against fires
  - ASR A2.3 Escape routes and emergency exits
  - ASR A3.4 Lighting and visual contact
  - ASR A3.5 Room temperature
  - ASR A3.6 Ventilation
  - ASR A4.2 Break and standby rooms
  - ASR A4.3 First aid rooms
  - ASR V3a.2 Setting up and operating workplaces
    National Committee for Occupational Safety and Health

- Guidelines for the Workplace Ordinance, LASI LV 40

  - National Committee for Occupational Safety and Health

- "Lighting of workplaces", LASI LV 41
- Screen Work Ordinance (BildscharbV)
- DIN 4543:1999: Office workplaces—Part 1: Areas for the installation and use of office furniture
- DIN EN 527-1:2008: Office furniture—Part 1 Office desks, Dimensions (draft standard)
- DIN EN ISO 9241-5:1999: Ergonomics requirements for office activities with visual display terminals
- DIN 16555:2002: Areas for communication workstations in office and administration buildings

- DIN 5035-7: Lighting of VDU workstations
- DIN 18041: Audibility in small to medium-sized rooms
- VDI 2058-3: Assessment of noise at the workplace taking into account different activities
- VDI 2569: Sound insulation and acoustic design in the office
- BGI 650: Bildschirm und Büroarbeitsplätze, Leitfaden für die Gestaltung, Verwaltungsberufsgenossenschaft VBG 2012 (German Association for Statutory Accident Insurance and Prevention in the Administrative Sector)
- BGI 5001: Office work—safe, healthy and successful, practical aids for design, VBG
- BGI 774: Office work system. Help with the systematic planning and furnishing of offices. VBG
- BGI 5050: Office space planning. Help for the systematic planning and design of offices. VBG 2005
- BGI 5128: Safe planning and design of workplaces. Guideline VBG 2008
- BGI 5019: Using buildings effectively; facility management solutions and practical aids for operators and users. VBG
- BGI 773: Call center. VBG

10.3   Reasons and Triggers for Diseases

Stress factors in the office (cf. Spath et al. 2011)

- One-sided postures
- Lack of exercise
- Pressure to meet deadlines and perform
- Disruptions and interruptions during the work activity
- Non-skilled or unskilled skills are required

Mental strain is caused by (ibid.):

- Overstraining activities
- Pressure to meet deadlines and assume responsibility
- Inappropriate room for maneuvers
- Conflicts with colleagues and business partners

Symptoms of mental illness (ibid.)

- The person appears indifferent or dismissive or aggressive.
- The person is subject to strong mood swings.
- The person closes herself off and isolates herself.
- The person shows decreasing performance or strong fluctuations in performance.
- The person does not dare to do anything anymore, seems insecure in general.
- The person takes many breaks and is remarkably often ill.

- The person feels bullied, personally assaulted, or attacking others.

Causes of mental illness (ibid.)

- Genetically predisposed
- Misguided biochemical processes in the body
- Nutrition
- Drugs
- Liquor
- Early childhood imprints
- The way in which a person perceives unavoidable risks, worries, and hardships in life and his feelings about them

Factors of msculoskeletal disorders (ibid.)

- External factors: Moist and cold climatic influences can contribute in the long term to muscle tension, local circulatory disorders, and immune deficiencies in the area of the back muscles as well as rheumatic diseases. Traumas and injuries also contribute to the development of MSE.
- Physical strain and tension: Lack of exercise or heavy physical work can lead acutely to painful muscle tension and mispositioning of the vertebral joints. Over a longer period of time, hard work, especially in connection with dissatisfaction, time pressure, one-sided strain and forced postures, promotes wear and deformation of the bony vertebral bodies and the intervertebral discs. MSE is also more common in overweight patients.
- Psychological factors: The posture of the spine expresses the inner posture. Stress, anger, and suppressed feelings such as anxiety and anger can lead to painful muscle tension. In the case of anxiety-induced body tension, pain is often experienced as a threat. In the long run, such factors favor the occurrence of chronic back problems, which can be accompanied by anatomical changes.

Burdening factors (ibid.)

- Lack of sleep and irregular sleep-wake rhythm
- Stimulating stimulants such as coffee and other caffeinated drinks, nicotine and sweets that counteract relaxation.
- Suppressed emotions such as anger, worry, anger and excitement, interpersonal conflicts and disharmonies.

Impairments in working hours at the monitor (cf. Ulich 2001)

- Four hours: Reduction of visual acuity and color sense disturbance (15–35 min regeneration time)
- Three hours: Visual acuity reduction, color sense disorders, and physical fatigue, eye fatigue (10–15 min regeneration time)
- Two hours: Accommodation and adaptation disorders, visual acuity reduction, and color sense disorders (15 min regeneration time)
- 1 h: visual acuity reduction and color sense disorders (10 min regeneration time)

According to Ulich (2001), **humane work tasks stand out**

- Holistic nature of the task in terms of planning, objective, execution, target/means decision, and control
- The scope for making decisions (i.e., the possibilities and requirements for making decisions)
- Diversity and variability of requirements
- Contact and communication requirements
- Transparency of the task context
- Scheduling of the day and order of activities to be performed according to their importance and urgency
- The mental willingness to perform is highest at around 11:00 o'clock in the morning. It is advisable to set demanding tasks for this point in time, as the ability to concentrate is particularly good here.
- From 12:00 o'clock on, the performance level decreases and the midday low begin. This time can preferably be used for telephone calls and short meetings.
- The lunch break must be observed regularly. After lunch, which should not be too lavish, a 20 min rest and relaxation break is recommended.
- The early afternoon is ideal for meetings and conferences.
- The second activity high of the day begins at 15:00. The long-term memory works particularly well and the motor skills are high.

The Following should be observed with regard to nutrition (ibid.)

- The height and quality of the grease supply
- The level and quality of carbohydrate intake
- Supply of calcium, iodine, fluoride, vitamin E, vitamin D, beta carotene, folic acid, and iron
- Sufficient fluid intake

Mental fitness (ibid.)

- Ability to concentrate and solve problems
- Attention
- Self-discipline

Sick building symptoms (Spath et al. 2011)

- Fatigue
- Sleep and sleep-through disorders
- Heaviness in the head
- Itching, burning of the eyes
- Headaches
- Irritated, stuffy, or runny nose

An office room must be (ibid.)

- Provide each employee with a work center.
- Support functional and social communication.
- Ensure undisturbed, concentrated work.
- To enable individual changes depending on the work requirements.
- Have a common infrastructure.

Recreation

- Autonomously regulated recovery: comprises unconsciously occurring processes of the vegetative and central nervous system that are activated on the basis of physiological regulation and protective mechanisms. Such protective mechanisms are partly deliberately controllable. One example of this is the use of stimulants to combat fatigue.
- As an emotionally regulating process: means that individual attributions of meaning have a decisive influence on recovery behavior.
- As a mentally regulated process: refers to the deliberate restoration of individual performance prerequisites. A well-known example is the regular observance of midday rest breaks for recreational purposes.

Break indicators

- Opportunity for recovery: Every break contributes in some way to recovery, so the argument of recovery is often used as a justification for breaks.
- Prevention of fatigue: Pauses can be used both to eliminate fatigue symptoms and to prevent fatigue. Both aspects cannot always be exactly distinguished in everyday working life. In addition to difficult physical working conditions, the prevention of fatigue plays a particularly important role in demanding mental work.
- Increased performance: breaks that prevent or compensate for fatigue also contribute to maintaining the health of the worker. By increasing performance, such breaks can also be worthwhile in the business sense. A rewarding break is defined as a work interruption in which the loss of performance during the break is compensated for by an increase in performance as a result of the recovery effect.
- Maintaining a sufficient level of vigilance: In activities characterized by constant information intake or low activity content, misuse leads to fatigue-like phenomena of monotony or saturation. An obvious symptom of such conditions is a lowering of the level of alertness, which can be eliminated by pausing or changing work.

Cause-effect chains for measures of a humane work design in the office (cf. Braun 2010)

- Economic benefit: Additional revenues and cost savings versus investment costs
- Productivity targets and customer orientation: strong customer orientation, high reliability, and demand-oriented solutions
- Service processes: Flexibility through self-organization, optimal process coordination, and reduced personnel requirements
- Employees: motivation and commitment, employee involvement, and low staff turnover
- Reduced disease risk
- Work design measures: task-oriented qualification, diverse work activities, and use of ergonomic work equipment

Dealing With Colleagues (Rules) (see Clariant 2015)
Dealing With Open Structures (Employees, Leadership)
We Build Mutual Trust.

- We treat each other respectfully and openly.
- We rate performance higher than the mere presence check.
- Control is good, trust is better.

We live appreciation and good behavior.

- We salute and treat each other friendly.
- We are considerate during informal meetings and breaks.

We treat offices and meeting rooms as if they were closed rooms when passing by.

- We do not look in there very hard.
- We rather greet here in exceptional cases.

   Conducting formal/informal meetings

We also hold meetings outside the meeting rooms.

- We use—especially for informal meetings—the cafeteria.

We book meeting rooms disciplined.

- The booked meeting room fits the size.
- Our booking takes place in time.
- If the meeting is omitted, we will immediately release the room again.
- If in doubt, we just ask.
- Can we swap meeting rooms?
- Can we reschedule the meeting to allow for both meetings?

Telephony and communication at work and in the home

Our communication is targeted.

- Our e-mails and messages are short and concise, the distribution list is well-considered.
- Alternatively—in complex situations—it is better to have a personal conversation.

Our communication behavior is characterized by consideration.

- The volume of our phone calls is reasonable.
- We only use loudspeakers if everyone in the room participates in the telephone call.
- Calls across multiple desks or floors should be avoided.

We maintain telephone discipline.

- In case of absence, we only call once and do not make multiple calls.
- We always leave a message, or we use e-mail when we want a call back.

Silent work (use of retreat rooms, free offices, etc.)

We value still work.

- We wll discuss it if anyone needs any rest.
- We give direct feedback if we feel disturbed.
- We show consideration and leave the office for intensive discussions when colleagues need to concentrate.

We go new ways for silent work.

- We can use free offices for quiet work after the previous and mutual arrangement.
- We also use the media library and seating groups for silence.

We avoid unnecessary disturbances.

- We close and open doors quietly.
- We make eye contact when entering a multi-person office and do not greet all those present loudly.

Order and cleanliness at the workplace and in public spaces

We treat our equipment with care at all times: meeting rooms, kitchenettes, sinks, tables

- Are left behind as we would like them to be ourselves.
- Are cared for by all of us. If necessary, we ourselves clear things away from others and remove coffee edges etc.

We put away dishes.

- If the dishwasher is empty, we put the dishes in the dishwasher. If it is full, we put the dishes on the plate above the dishwasher—not in the sink.

We indicate our private food in the refrigerator
with a piece of paper or by packing in cans—so we can quickly identify the owners.
Dealing with customers in-house
We arrange customer visits in the sense of mutual consideration.

- We explain basic rules to our visitors in advance and demand their observance.
- We consider the break times of the employees.

Our visitors and employees are at eye level.

- If possible, we will announce a visit on site to a laboratory or in analytics.
- In a direct meeting with an employee, we greet the employee and introduce him or her.

What applies to our employees also applies to our visitors.

- Smoking is only permitted in the designated areas.
- Meeting rooms are not workplaces - the cafeteria can be used for 1-2 h.
- Photography is prohibited.
  ... and in the end always applies: tolerance and feedback.

Literature

Bachmann, P., & Lange, M. (2013). *Healthy building with safety* (2nd en.). Springer Vieweg.

Braun, M. (2010). Promoting the ability to change within a company, designing work for people. *Safety Engineer, 4,* 8–15.

Clariant. (2015). *Dealing with colleagues (rules of the game)*. Frankfurt, Hoechst.

Heidenberger, B. (2017). *Time blossoms: time management, target management, working methodology ergonomics checklist,* www.zeitblueten.com/news/wohlfuehl-buero. Access on....

Kratzer, N. (2017). *"Open Ace. Oder was?" Prägewelt, rientierte Gestaltung neuer (Open Space) Arbeitswelten.* ISF München.

Martin, P. (2006). *Mobile office work—designing new forms of work humanely.* Düsseldorf: Hans Böckler Foundation.

Reimersdahl, Anke van. (2017). *The dear colleagues "Krümelmonster" and "Vetter It"*, https://bruynzeel-storage.com/de/hygiene-im-buero/ Access on....

Scratch, N. (2017). *"Open Space. Or what?" Prägewelt, prevention-oriented design of new (open space) working worlds.* ISF Munich.

Spath, D., Bauer, W., & Braun, M. (2011). *Healthy and successful office work.* Berlin: Erich Schmidt.

Steelcase. (2008). *Well-being: A theme that only winners know,* https://www.steelcase.com/eu-de/forschung/artikel/themen/arbeitsplatz/wohlbefinden-ein-thema-das-nur-gewinner-kennt/. Access on....

Ulich, E. (2001). *Industrial psychology* (5th ed.). Stuttgart: Schäffer-Poeschel.

Weber, L. (2017, MAy 23). Pale haze in the office. In: *Frankfurter Allgemeine Zeitung, Technology,* p. 0T1.

Index

© Springer Nature Switzerland AG 2020
W. Seiferlein and C. Kohlert (eds.), *The Networked Health-Relevant
Factors for Office Buildings*, https://doi.org/10.1007/978-3-030-22022-8

Printed in the United States
by Bookmasters

Printed in the United States
By Bookmasters